The Making of Irish Geology 1740–1940

Ireland has long been the focus of research into its geological foundations and history. This book provides insights into the evolution of geological ideas in Ireland, gives assessments of underrepresented scientists, and offers an appreciation of the value of geological collections for historical research.

Through a series of twenty-two papers, this volume provides syntheses and analyses of aspects of this geological research and places the role of a number of individuals firmly within context. The topics discussed include the considerations of the nature of igneous rocks, aspects of geological mapping of the country, the role played by Irish scientists in determining the age of the Earth, and the importance of early mineralogical and geological collectors and the role their collections made to advancing geological knowledge both in Ireland and Europe.

The Making of Irish Geology 1740–1940 will appeal to both the general and academic reader interested in the development of the geological sciences, with a particular focus on Ireland.

Patrick N. Wyse Jackson is a Professor in Geology and Fellow at Trinity College Dublin and a Member of the Royal Irish Academy. His work in the history of geology has focused largely on Ireland and on biographical studies of the major contributors to the science. He has published widely in national and international journals; his books include *The Chronologers' Quest: The Search for the Age of the Earth* (2006) and the edited volume *Four Centuries of Geological Travel* (2007). He is a member of the International Commission on the History of Geological Sciences (INHIGEO).

Variorum Collected Studies

Also in the Variorum Collected Studies series:

EVELLEEN RICHARDS
Ideology and Evolution in Nineteenth Century Britain
Embryos, Monsters, and Racial and Gendered Others in the Making of Evolutionary Theory and Culture (CS1089)

DAVID S. BACHRACH
Administration and Organization of War in Thirteenth-Century England (CS1088)

GÉRARD GOUIRAN, edited by Linda M. Paterson
From Chanson de Geste to Epic Chronicle
Medieval Occitan Poetry of War (CS1087)

JOHN A. COTSONIS
The Religious Figural Imagery of Byzantine Lead Seals II
Studies on the Images of the Saints and on Personal Piety (CS1086)

JOHN A. COTSONIS
The Religious Figural Imagery of Byzantine Lead Seals I
Studies on the Image of Christ, the Virgin and Narrative Scenes (CS1085)

WENDY DAVIES
Christian Spain and Portugal in the Early Middle Ages
Texts and Societies (CS1084)

PEREGRINE HORDEN and NICHOLAS PURCELL
The Boundless Sea
Writing Mediterranean History (CS1083)

MOHAMED EL MANSOUR
The Power of Islam in Morocco
Historical and Anthropological Perspectives (CS1082)

JOHN MARSHALL, edited by Philip Butterworth
Early English Performance: Medieval Plays and Robin Hood Games
Shifting Paradigms in Early English Drama Studies (CS1081)

JENNIFER O'REILLY, edited by CAROL A. FARR and ELIZABETH MULLINS
Early Medieval Text and Image II
The Codex Amiatinus, the Book of Kells and Anglo-Saxon Art (CS1080)

https://www.routledge.com/Variorum-Collected-Studies/book-series/VARIORUM

The Making of Irish Geology 1740–1940

Patrick N. Wyse Jackson

VARIORUM COLLECTED STUDIES CS1131

Routledge
Taylor & Francis Group
LONDON AND NEW YORK

First published 2025
by Routledge
4 Park Square, Milton Park, Abingdon, Oxon OX14 4RN

and by Routledge
605 Third Avenue, New York, NY 10158

Routledge is an imprint of the Taylor & Francis Group, an informa business

British Library Cataloguing-in-Publication Data
A catalogue record for this book is available from the British Library

ISBN: 978-1-032-75768-1 (hbk)
ISBN: 978-1-032-75769-8 (pbk)
ISBN: 978-1-003-47555-2 (ebk)

DOI: 10.4324/9781003475552

Typeset in Times New Roman
by codeMantra

VARIORUM COLLECTED STUDIES CS1131

CONTENTS

ACKNOWLEDGEMENTS

Over the last three and a half decades, my research within the field of the history of geosciences has benefitted from the advice and encouragement of many people. I owe a particular debt of gratitude to many valued friends in this regard. In particular, Gordon Herries Davies, Nigel Monaghan, Matthew Parkes, Hugh Torrens, and Ezio Vaccari have always been most generous with their time, and their collaborative openness and willingness to impart information where my research paths have coalesced with their own have allowed me into avenues that I might have not entered.

Many others have aided my research by providing information, insights, and advice, or by researching archives and geological collections in their care, and giving me permission to use this material. The editors and reviewers of these chapters when first submitted made insightful comments that directed me to make alterations and this aided their appearance and content on publication. In addition, I am most grateful to the publishers and my co-authors for granting permission to reissue chapters in this compilation. The summary of the geological work by J.W. Pringle in Chapter 8 is somewhat derivative of the 1983 publication of Gordon Herries Davies on geological mapping in Ireland.

I am most grateful to the following, some of whom have sadly passed away but are not forgotten:

Jean Archer
Claire Bailey-Coombs
Lawrence Badash
Heather Baily
Ossie Baily
Síghle Bhreathnach-Lynch
Ken Bork
Marie Bourke
William Brice
Stephen Brush
Ruth Buckley
Cynthia Burek
Declan Burke
Caroline Buttler
Duncan Thorburn Burns
Petra Coffey
Rhoda Coghill
Ann Cole
Colin Cole
Stephen Coonan
Con Costello
Gordon Craig
Mike Crane
Peter Crowther
Miguel De Arce
John Diemer
Anne Dingsdale
Philip Doughty

Frank Faughnan
Lydia Ferguson
Paul Ferguson
Sylvia Figueirôa
Delwyn Geraghty
Sibdas Ghosh
Rob Going
Greg Good
Andrew Grout
Beryl Hamilton
David Harper
Catherine Harpham
Duncan Hawley
John Henry
Sandra Herbert
Gordon Herries Davies
Bettie Higgs
Kathleen Histon
Charles Holland
Kenneth James
Andrew Jeram
Brónach Joyce
Padraig Kennan
Marcus Key
Maria Key
Simon Knell
Léo Laporte
Cherry Lewis
Colain MacArthur
Niall MacCaughey
Peadar McArdle
Frank McGarry
Graham McKenna
Daniel Merriam
Thomas Mitchell
Nigel Monaghan
Louise Morgan
John Murray
Sally Newcomb
Ída Ní Thuama
Ted Nield
John Nudds
Geraldine O'Connor
Siobhan O'Rafferty
David Oldroyd
Naomi Oreskes
Leslie Overstreet
Matthew Parkes
Beryl Phair
Earl of Ranfurly
Raymond Refaussé
Wesley Riddle
Eric Robinson
Ted Rose
Patrick Roycroft
Martin Rudwick
Christine Ryall
Sunil Sen Sarma
Ann Scanlon
Susan Scarcella
John Schaffter
Len Scott
George Sevastopulo
Brian Shipley
Tom Sharpe
Mike Simms
Mary Spencer Jones
Nicholas Stanley-Price
Adam Stuart Smith
Brian Taylor
Ken Taylor
Mike Taylor
Paul Taylor
John Thackray
Hugh Torrens
Ezio Vaccari
Denis Weaire
Gerard Whelan
Leonard Wilson
Tony Wright
John Wyse Jackson
Robert Wyse Jackson
Ellis Yochelson

and I owe my largest thanks to my family Vanessa, Susanna, and Katie who have journeyed far and wide with me and the litany of characters discussed in this book.

LIST OF FIGURES

INTRODUCTION

Ireland, despite being situated at the western edge of Europe, has long been the focus of research into its geological foundations and history. Much was significant and fed into a wider understanding of geological processes and the Earth's history and chronology. Aside from a number of scholarly assessments of the mapping of Ireland's rocks, much of this pioneering research had not until recently been the focus of attention.

This volume focuses on the development of the understanding of Ireland's geological foundations and its history over a period of 200 years. A strong thread interwoven running through this volume, and the overarching rational for this research, is to firmly examine this development through the lens of individual scientists and savants and Irish institutions and to place their research in context. Topics discussed include considerations of the nature of igneous rocks, aspects of geological mapping of the country, the role played by Irish scientists in determining the age of the Earth, and the importance of early mineralogical and geological collectors and the role their collections made to advancing geological knowledge both in Ireland and Europe.

Just as there is a closer relationship between the army and some of the founding fathers of British geology than is commonly realised (Rose, 1996, p. 129), so there is a closer relationship between geological work in Ireland and the development of major interpretative ideas in earth science than is generally appreciated. The resolution of the eighteenth-century debate on the nature and origin of various igneous rocks, including basalt, was partially reached through work carried out by several geologists in north-east Ireland (Herries Davies, 1981; Wyse Jackson, 2000). Moreover, Patrick Ganly publishing in 1856 showed how sedimentary structures could be used to determine way-up in rocks (Archer, 1980; Wyse Jackson, 1995), and Joseph Beete Jukes' work in 1862 on the river drainage of southern Ireland was highly influential in shaping the thoughts of later geologists and geomorphologists on river drainage and its effects on the landscape in general (Herries Davies, 1962). Geological work in Ireland extends back over 300 years (Wyse Jackson 2001b). Its significance is being increasingly documented by local historians for the benefit of the international scientific community (see Wyse Jackson, 1997a [Chapter 1 herein] for a comprehensive review of this work).

 DOI: 10.4324/9781003475552-1

The tripartite theme adopted in this volume is firstly to show that the study of Ireland's geology not only was important for the development of the Irish and international geological community and the evolution of geological ideas and theories but also was central to aims of various Irish-based institutions and learned societies who viewed geology as a subject worthwhile pursuing and promoting.

Secondly, this book demonstrates the value of collective biography in that it brings to focus the work of lesser-known geological and mineralogical practitioners such as the clerics William Richardson who championed ideas on the formation of the Giant's Causeway in north-east Ireland, and George Graydon who was an early traveller and collector to the volcanic regions of Italy, as well as the Honourable George Knox, an Irish member of the nobility and parliamentarian, who had a deep passion for minerals. It evaluates some early methods of mapping offshore geology and highlights the geochronological estimates made by John Joly and William Thomson, Lord Kelvin. Discussed are those like William Fitton who advocated for the work of others, or those like John Pringle who acted administratively to promote geological mapping and surveying. A number of studies focus on individuals who advanced the development of new avenues for research such as Grenville Cole who championed in Ireland the study of landscape or the earlier inspired mapping of marine geology by Alexander Nimmo. John Joly, perhaps Ireland's greatest scientific polymath, delved into geochronology but didn't make the leap in appreciating how radioactive minerals could provide absolute geological dates.

Beyond Ireland's shores, valuable contributions were made to the understanding of the geological rocks and structure of Italy and India, thanks to the travels and permanent translation of Irish geologists. At home, shorter expeditions and fieldwork campaigns traversed Ireland's bogs and mountains and coastlines, while other experimentalists remained rather static within a dank laboratory, institution, or even in a domestic setting. All, nevertheless, added to the growing volume of literature pertaining to Ireland's geology.

Within these pages, various individuals, some of whom remained hidden in time or largely forgotten, are brought to the forefront. The geological endeavours of William Brownrigg and Patrick Ganly were underacknowledged in their lifetimes, while the pioneering research of a number of female geologists such as Sydney Thomson is only now getting more widespread appreciation by a number of modern commentators. Their research in Ireland was critical in glacial dynamics, igneous petrology, and Silurian stratigraphy. We can now understand better the roles of these individuals and their circles and appreciate that some had an international influence—for example, activities in Ireland and by Irish researchers were central to the debate that underpinned the early considerations of the nature of basalt and lava.

Thirdly, this research demonstrates the value of geological collections in developing a clear understanding of the development of geological theories. While archival studies are normally the remit of most historians of geology, close examination of geological specimens residing in museums together with parallel considerations of archival materials, catalogues, letters, and the printed output of early scientists can be most illuminating. For example, the modern identity of 'blisterstone' described in the early 1800s is revealed through the examination of museum samples to be an amygdaloidal basalt.

Why 1740–1940? This is not to suggest that interest and research on Ireland's geology didn't occur before the former date nor continue in the eighty-plus years after 1940. The year 1740 marked the period when the Giant's Causeway became generally known to scientific travellers from abroad, while post-1940 saw the initial expansion of both Government-funded geological surveys and the university departments. Their subsequent history has partly been told by Herries Davies (1995) and Wyse Jackson (1994b, c), but it is up to a future geohistorian to assess the impact of geological research on the island from this date to the present.

This volume draws together twenty-two papers published over three decades that raise the profile of Ireland's geological pioneers and demonstrate the value of geological materials for historical research. This selection of publications taken together allows for a deeper understanding of the evolution of geological ideas in Ireland. This is a personal choice of the contributions of many scientists, persons of thought, and savants of natural history, but others could have been included in the pantheon. To name only one, I think of Samuel Haughton, but I am sure that readers will bring others to mind.

In the papers reproduced here some modifications have been made to the original published versions where errors or omissions on my part were kindly highlighted by friends, while in some papers a small amount of additional information is added from relevant other publications of mine. Some citations to more recent valuable research on the history of Irish geology have been added, particularly to the Chapter 1.

REVIEW OF THE HISTORY OF GEOLOGY IN IRELAND

1

FLUCTUATIONS IN FORTUNE: THREE HUNDRED YEARS OF IRISH GEOLOGY 1694–1994

In J. Wilson Foster and H. Chesney (eds.) *Nature in Ireland.* Lilliput Press, Dublin (1997), pp. 91–114.

Introduction

The origins of geology as a science in Europe lie in the early eighteenth century, when it developed out of the widespread interest in natural history, particularly in minerals and mining. From the 1600s, mining methods became more advanced and came to be widely known across Europe through the writings of Agricola and others. In Germany, institutions were established to promote scientific thought, and mineralogical studies advanced rapidly from the 1700s onwards. In these islands, the first formal societies that promoted science were established in the mid-1600s, with the Royal Society (1660) and the Oxford Philosophical Society (1683) in England and the Dublin Philosophical Society (1683) in Ireland.

The development of Irish geology owes as much to the collective efforts of various learned societies, government agencies, and universities as to the individual men, and latterly women, who have carried their hammers and packs across Ireland's countryside. While a number of societies and groups such as the Dublin Philosophical Society, the Physico-Historical Society, and the Geological Society of Dublin/Royal Geological Society of Ireland—have come and gone, others, including the Royal Irish Academy, the Royal Dublin Society, the Geological Survey of Ireland, and Trinity College, Dublin, have remained active in geology for over 170 years.

Early geological investigations in Ireland, 1683–1750

In Ireland, the first formal geological investigations were carried out by members of the Dublin Philosophical Society, 1683–1708 (Hoppen, 1970). William

DOI: 10.4324/9781003475552-3

Molyneux (Gabbey, 1985) was the main force behind the establishment of this society, where he was joined by men such as the cartographer Sir William Petty and the anatomist Allen Mullen (Kelly, 1992). Through papers published in *Philosophical Transactions of the Royal Society of London*, various aspects of Ireland's geology were examined by members of the Dublin Philosophical Society. Probably for the first time, two gems of Irish geology were described in print: the Giant's Causeway by the Rev. Samuel Foley and Thomas Molyneux (brother of William) in 1694 and the Giant Irish deer by Thomas Molyneux in 1697.

In 1744, the Physico-Historical Society of Ireland was established in Dublin to survey the natural resources of the country (Herries Davies, 1979). It employed several people, including Charles Smith and the botanist Isaac Butler, who were to survey each county, collect botanical and geological specimens, and write an account of their findings. Reports appeared for only four counties: Cork, Down, Kerry, and Waterford. The Cork volume by Smith was accompanied by a map that portrayed the east-west trending synclinal valleys floored with limestone. It was the first attempt to depict the solid geology of any portion of Ireland (Herries Davies, 1983). Though the Physico-Historical Society of Ireland lasted only a dozen years, many of its aims were later pursued by the Dublin Society.

Development of mineralogical and geological studies, 1750–1825

From the mid-1700s, mineralogy became a focus for serious scientific study, and from it, the science of geology developed (Laudan, 1987). Ireland, and in particular Dublin, became an important centre for mineralogical and geological studies, ranking close to centres of excellence such as Freiburg and Edinburgh. There were three main reasons for this: the establishment of various institutions that promoted these studies, such as the Royal Irish Academy, the Dublin Society (later the Royal Dublin Society), and the Belfast Literary Society; the international stature of Irish scientists; and the debate on the origin of basalts, centred on the Antrim coast, which was of fundamental importance to the development of geological thought in Europe.

The Royal Irish Academy was founded in 1785 for the study of sciences, 'polite literature' and antiquities (O'Raifeartaigh, 1985). It elected its members from the noble, clerical, and academic circles of the day and rapidly built up an important library and museum. Geological problems were among those tackled by members (Herries Davies, 1985a), and many papers on such topics were published in the Academy's *Transactions*. This journal was published from 1798 until 1907 and copies were sent to similar learned institutions in Europe, thus allowing for a rapid diffusion and exchange of ideas (Herries Davies, 1978b). The Academy also developed a museum that contained important archaeological material and some geological specimens and became the official repository for the treasure trove found in Ireland. Many years later, the geological collections were incorporated into the collections of the Dublin Society; they are now in the National Museum of Ireland.

Perhaps the most influential member of the geological fraternity at the time was Richard Kirwan, a Galway-born chemist and geologist and President of the Academy from 1799 to 1812 (McLaughlin, 1939–1940; Burns, 1985; Wyse Jackson, 1998c [Chapter 4 herein]). Through his book *Elements of Mineralogy* (1784), he became one of the most influential mineralogists of his day. In it, he coined the term 'Calp' for a flinty dark grey to black limestone quarried near Dublin. In later life, Kirwan became a noted eccentric who disliked flies so much that he paid his servants for each corpse presented to him. He also disliked replying to correspondence and had his door-knocker removed each evening at 7 o'clock to deter further visitors. He died in Dublin while engaged in the practice, common for the time, of starving a cold (Somerville-Large, 1975).

Within the pages of the *Transactions* were published a suite of papers dealing with a controversy that began at the end of the eighteenth century: the origin of granite, basalt, and other igneous rocks. One school of thought, the 'Neptunists' or, to use a contemporary term, 'watermen', believed that these rocks were precipitated from water—a view held by the German mineralogist Abraham Gottlob Werner and his students in the Freiburg School of Mines. Kirwan also followed the theories of the Neptunists. The other group, variably known as 'Vulcanists' or 'firemen', according to the rock types in which they were interested, argued that the rocks were the products of volcanoes and other igneous phenomena (Herries Davies, 1981). These men included the Rev. William Hamilton, a Fellow of Trinity College, Dublin, who wrote an important and influential memoir entitled *Letters Concerning the Northern Coast of the County of Antrim* (Dublin, 1786). In 1790, he became rector and a local magistrate of Clondavaddog, an isolated parish in Co. Donegal, but was brutally murdered seven years later after local unrest which culminated in the uprising of 1798. The debate was centred around the basaltic Giant's Causeway in County Antrim, which made Ireland one of the more important sites for resolving it. In the late eighteenth and early nineteenth centuries, the Giant's Causeway drew many visitors from Britain and some from continental Europe, including John Wesley in 1778, John Whitehurst in 1783, Abraham Mills in 1787–1788, Humphry Davy in 1806, and Jean-François Berger in 1811 (Herries Davies, 1978a; Siegfried and Dott, 1980; MacArthur, 1990; Wyse Jackson, 2000 [Chapter 2 herein]). Its unusual character had been widely advertised by Frederick Hervey, Bishop of Derry and fourth Earl of Bristol, and by Susanna Drury, who won a commission of £25 from the Dublin Society in 1740 to paint views of the Causeway. Her gouaches were subsequently engraved and distributed widely, thus bringing the area to the attention of many travellers and scientists (Anglesea and Preston, 1980; Anglesea, 1997). Both factions in this geological conundrum felt that they could explain the formation of the basalts according to their principles. The Neptunist or Wernerian view was reinforced by the presence of marine shells called ammonites thought to be in basalt occurring at Portrush, County Antrim which were first described in 1803 by the Rev. William Richardson (Blackstock, 2013), a vociferous 'waterman' (Herries Davies, 1985a). However, a detailed examination of the ammonites by Sir James Hall and

John Playfair in Edinburgh revealed that they were present in a sedimentary rock that had been baked by hot basalt close by, giving it a similar appearance to that basalt (Playfair, 1802). These conclusions were confirmed by the Rev. William Daniel Conybeare and Jean-François Berger in 1816. This piece of evidence was critical and swung the scales in the favour of the Vulcanists. Additional evidence came from observations made in Italy of active and extinct volcanoes—especially Vesuvius, which had erupted in the 1790s—and from a suite of Italian volcanic specimens presented to the Museum of the Academy by one of its members, the Rev. George Graydon, who had personally collected them so as to demonstrate the similarities between Antrim and the Italian areas (Wyse Jackson and Vaccari, 1993; Vaccari and Wyse Jackson, 1995 [Chapter 3 herein]). Another paper of note was that by a Scottish civil engineer, Alexander Nimmo, which described the geology of offshore Ireland (Nimmo, 1825a, b; Wyse Jackson, 1996a [Chapter 7 herein]). Some 140 years later, this area was to become well known, thanks to feverish exploration for oil and gas. Nimmo made significant contributions to the opening up of the west of Ireland by constructing bridges, roads, and harbours (Villiers-Tuthill, 2006) and also carried out survey of Irish bogs.

The Dublin Society and other institutions

The Dublin Society was established in 1731 to promote agriculture and sciences in Ireland. It did so by offering premiums to members to carry out research, organising county statistical surveys, acquiring specimens for its museum, and acting as a forum for discussion, debate, and study (Berry, 1915; Meenan and Clarke, 1981; Mollan, 1990; Bright, 2004). In 1786, Donald Stewart, appointed as Itinerant Mineralogist, studied the geology of various parts of the country and collected specimens for the Society's museum (Stewart, 1799). The museum's large and important collection contained 7,331 mineral specimens assembled by the German mineralogist Nathanael Gottfried Leske (1751–1786), which had been purchased in 1792, largely thanks to the efforts of Kirwan, for the huge sum of £1,350 (the equivalent of approximately £35,000 today) (Vaccari and Monaghan, 1993). The acquisition remains geologically the most important ever made by an Irish museum. It stimulated considerable geological activity by members who made mineral collections for themselves, many of which ultimately found their way into the museum or other institutions. It also encouraged geological research and one of the highlights was a pamphlet by John Hart who described in detail the occurrence and physiology of the Giant Irish Deer (Hart, 1825, 1830; Wyse Jackson, 2019a; Figure 1.1).

Between 1801 and 1832, the Dublin Society published the results of county statistical surveys of twenty-three counties, each of which gives some geological information. Of particular note are the volumes on Wicklow (1801) by Robert Fraser, Kilkenny (1802) by William Tighe, Londonderry (1802) by George Vaughan Sampson, and Cork (1810) by Horatio Townsend, all of which contained highly advanced geological maps (Herries Davies, 1983).

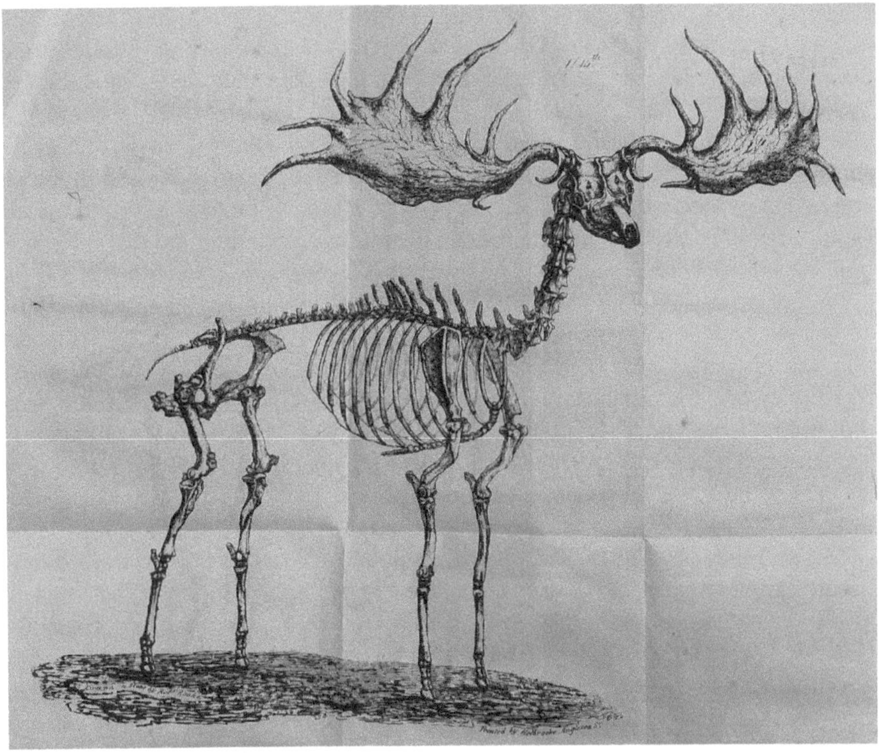

Figure 1.1 Giant Irish Deer as reconstructed by John Hart in 1825 for the Royal Dublin Society (Geological Museum, Trinity College, Dublin).

Richard Griffith, who went on to make a profound contribution to Irish geology (Herries Davies and Mollan, 1980; Wyse Jackson, 2008a [Chapter 9 herein]), was engaged to report on the Leinster coalfield in 1809 and was appointed as Mining Engineer in 1815. He surveyed and mapped various other coalfields, including those of Connacht (1818) and counties Tyrone and Antrim (1820). The Dublin Society attempted to instigate a geological mapping scheme for the whole country, but this did not happen until it was officially carried out by the Geological Survey, and unofficially by Griffith.

In order to promote geology further, the Dublin Society decided in 1812 to appoint a Professor of Mineralogy. The position was first offered to Robert Jameson of Edinburgh, who had visited Dublin in 1797 to study the Leskean collection (Sweet, 1967). Jameson accepted the post but later rescinded his acceptance, so the job went to the German-born mineralogist Sir Charles Lewis Giesecke. He is best known for his collection of minerals acquired in Greenland in the early years of the 1800s (Praeger, 1949), for which he was knighted by the Danish king, and for appearing in the first performance of Mozart's *The Magic Flute* in Vienna in 1791 (Whittaker, 1991). Many of his Greenland specimens were hijacked at sea and eventually ended up in Scotland. Others were deposited in Copenhagen, many were donated to the Dublin Society Museum (Monaghan, 1993), and some have been recognised in the collections of Trinity College, Dublin (Wyse Jackson, 1996b [Chapter 22 herein]). Giesecke rapidly learned English so that he could lecture on mineralogy. He catalogued the collections (Giesecke, 1832a, b) and made several collecting trips around Ireland (Giesecke, 1825a, b, 1826, 1828, 1829), adding many specimens to the Dublin Society Museum. Giesecke also arranged for material to be exchanged with foreign institutions, including the Royal Institution, Vienna, and the German Goethe Institute (Hennig, 1951; Waterhouse, 1933).

In Cork from 1790, Thomas Dix Hincks, a Unitarian minister, instigated lectures on 'general and useful knowledge'. As a result of this work, the Royal Cork Institution was founded in 1802 (Pettit, 1976). It soon ran a botanic garden and employed four professors, one of whom lectured in natural history, and another of whom was the accomplished chemist Edmund Davy, a nephew of Sir Humphry, who himself had collected geological specimens in Co. Donegal in 1805 and 1806 for use in his lectures at the Royal Institution in London (Davis, 1985; MacArthur, 1987). The Royal Cork Institution was later assimilated into the Queen's College, Cork.

During this period in Belfast, a small number of learned bodies were established that rapidly became the cultural focus for the north-east part of the country. These never attained the recognition accorded their Dublin counterparts, probably because their membership was drawn from the middle and working classes, but nevertheless, they were important in the development of the study of geology in that city. The Belfast Reading Society (later the Belfast Library and Society for Promoting Knowledge, properly known as the Linen Hall Library), which was founded in 1788, possessed a collection of fossils. The Belfast Literary Society, founded in 1801 (Anon., 1902), also kept a small cabinet of minerals, many of which were possibly collected by its first president, James McDonnell, who was the geological mainstay of the society and who read several papers to it (Doughty, 1980). Both of these small collections were passed on to the Belfast Natural History Society in the 1830s. This society was established in 1821 by James Lawson Drummond in association with others, including the zoologist Robert Patterson and Hincks of the Royal Cork Institution (Deane, 1924; Nesbitt, 1979). Soon collections of geological and other materials were made, and these laid the foundations for the present Ulster Museum. In 1831, James Bryce, a schoolmaster in Belfast Academy, published a catalogue of their collections, which were assembled for 'the cultivation of the science generally' (Bryce, 1831). The work of the Belfast organisations played an important role in fostering an interest in the natural sciences among the population of a city where few people had access to formal education (Nesbitt, 1979).

Ordnance Survey and geological survey mapping

From the 1820s onwards, for over six decades, Ireland was an important centre of scientific research through the work of men such as William Rowan Hamilton, the mathematician and inventor of quaternions; the physicist Humphrey Lloyd; the astronomers William Parsons, third Earl of Rosse, who built a telescope with a 72-inch reflector at Birr, Co. Offaly, and Robert Stawell Ball, the Astronomer Royal; the seismologist Robert Mallet; and the geologists Rev. Samuel Haughton and Sir Richard Griffith (Wyse Jackson and Wyse Jackson, 1992). Geological study in Ireland flourished: primary mapping of the solid geology of the country was initiated and completed (Herries Davies, 1995); a strong and influential geological society was very active; geology was formally recognised as a university subject; and many of Ireland's geologists, professional and amateur, attained a high international profile.

In 1809, the Commissioners of Bogs began their mammoth task of surveying Ireland's boglands. Soon afterwards, in 1824, the Ordnance Survey under Lieutenant-Colonel Thomas Colby began to survey the country and produced the first large-scale topographical maps. The scene was set for geological and economic mapping of the country. John W. Pringle was brought in to oversee geological activities and produced a written outline of instructions for geological and mineralogical observations (Rose, 1996, 1999; Wyse Jackson, 1997b [see Chapter 8 herein]). At first, the Ordnance Survey staff collected topographical and demographical information, as well as geological specimens and observations (Herries Davies, 1974). It published a volume entitled *Memoir of the City and North Western Liberties of Londonderry. Parish of Templemore* (1837), which included a geological map. Unfortunately, a collection of information for such a memoir stretched resources, and no further parish reports appeared. However, another report emanated from the Ordnance Survey when Joseph Ellison Portlock, together with Thomas Oldham and a staff of more than thirty, produced in 1843 a map and a bulky memoir entitled Report on the Geology of the County of Londonderry, and of Parts of Tyrone and Fermanagh (Seymour, 1944). The geological data collected during this work was for the first time plotted on the large-scale six-inches-to-the-mile maps which had only just been published. Portlock was an engineer who spent considerable time in Ireland on the basic measurement or triangulation of the country, very often in harsh conditions. In 1828, he overwintered 1,200 feet up on Slieve League. For his efforts, Portlock was rewarded by being sent to Corfu, where he supervised the installation of fortifications. He remained there until 1849, after which he spent some time in Britain before retiring to Dublin in 1862. Portlock was a highly regarded member of the geological fraternity; he was a Fellow of the Royal Society and President of the Geological Society of Dublin and also became a Major General in the army. Today, his palaeontological collections, many of which were collected by Patrick Doran of Glassdrummond, and by James Flanagan, are largely housed in the British Geological Survey and the Ulster Museum (Tunnicliff, 1980) and remain an important research asset for modern geologists. While Flanagan continued to earn his salary collecting fossils for the later Geological Survey, Doran became one of the first itinerant mineral collectors and dealers in Ireland. Many of his specimens found their way into collections, including those of the Royal Dublin Society, the Belfast Natural History and Philosophical Society, and the British Geological Survey (Cleevely, 1983).

Two years after the publication of Portlock's Londonderry memoir, the Geological Survey was established under the general directorship of Thomas De la Beche and the local directorship of Henry James. For the next forty-five years, the officers of the Geological Survey in Ireland were engaged in producing a geological map of the country. Their first published maps depicted the geology of various counties at a scale of one inch to two miles. The geological colours that indicate the various rock types were hand-painted onto base maps of the Index sheets to the Townland Series. Five such maps appeared: Wicklow (1848), Carlow (1849), Kildare (1849), Wexford (1851), and Dublin (1851), this last at a scale of one inch to one and a half miles. The problem facing the geologists was that there were no maps at the preferred scale of one inch to one mile. They had to wait a number of years before such maps became available from the Ordnance Survey. The first geological map at this scale was published in 1856, accompanied by a memoir that contained an account of the geology depicted on the map. The Geological Survey was to produce a total of 205 such map sheets and memoirs before its task was completed (Herries Davies, 1983, 1995). The importance of the Geological Survey to Irish geology at this time cannot be overstated. In men such as Edward Hull, Joseph Beete Jukes, George Henry Kinahan, George Victor Du Noyer, and Thomas Oldham, it employed some of the most eminent geologists to work in Ireland. Du Noyer was a fine geological artist, and his field sheets and notebooks are liberally adorned with quick but perceptive sketches (Archer, 1993, 1995; Coffey, 1993; Croke, 1995; Monaghan, 2001, 2017; Hegarty, 2017). Oldham did not remain in charge of the Survey for very long, and in 1850 he removed himself and his new wife to India to take charge of the new Geological Survey of India, based in Calcutta. Over the next twenty years, he succeeded in persuading many Irish geologists to join him there (see Wyse Jackson, 2008b [Chapter 13 herein]).

Oldham was replaced by Jukes, an Englishman born near Birmingham who appointed an additional ten men to the Geological Survey and began an ambitious regime of field mapping. From 1850 to 1870, an average of about a thousand square miles was covered per year, and from 1856, the one-inch-to-one-mile maps were published. By 1860, the maps for most of Munster and south-east Leinster were published, by 1869 maps of all of Leinster and east Galway were available, and by 1890 geological mapping of the whole country had been completed.

The mapping programme did not always proceed smoothly, however; there were frequent clashes of personality and opinion. The most serious of these was that between Hull, who had replaced Jukes as Local Director in 1869, and Kinahan, who resented Hull's appointment. This infighting led to numerous delays in the publication of maps and memoirs. In addition, Hull's revisionist interpretation of some of the geology of southern Ireland was questioned by his team, and subsequently, in the period 1903–1912, the areas affected by Hull's work were resurveyed.

The conditions endured by the geologists were not always pleasant. Salaries were low: Jukes earned £575 as Director in 1865, while an Assistant Geologist was paid twelve shillings per day (Herries Davies, 1983). In addition, the geologists mapped in typical wet Irish weather and moved from field station to field station, which did not allow them to put down roots. In some cases, the adverse conditions led to a breakdown in health, and a few geologists actually succumbed while surveying: Frederick Foot and William Benjamin Leonard were drowned; Frederick William Egan died after being thrown from a car; Jukes, while in Kenmare on a tour of inspection, fell down a flight of stairs and never fully recovered; Joseph O'Kelly died of bronchitis; and Du Noyer died of scarlet fever in 1869. Altogether, of the thirty-two men who worked for the Geological Survey between 1845 and 1890, seven died while in its employment (Herries Davies, 1983).

The end result of the work of the Geological Survey was the publication of a series of maps of the highest standard which for many parts of Ireland remained until very recently the only available geological maps. The one-inch-to-one-mile geological map of Ireland is a lasting testament to the labours of the thirty-two geologists employed by the Geological Survey of Ireland in this period.

Richard Griffith's unofficial geological survey of Ireland

From the 1830s, Richard Griffith was engaged in compiling a geological map of the country. Griffith, who had studied under the eminent mineralogist Robert Jameson in Edinburgh, came under the influence of George Bellas Greenough, who encouraged him to start this task. Griffith had started collecting geological information during his period of employment with the Dublin Society and as Engineer to the Office of Public Works. His geological activity received a boost in 1830 when he was appointed Commissioner for the General Valuation of Rateable Property and as such he became a noted public figure and received a baronetcy in 1858 (Wyse Jackson, 2008a [Chapter 9 herein]). For the Valuation, he employed many surveyors, who were instructed unofficially to make geological observations. Indeed, Patrick Ganly, one such employee, spent almost all his time mapping geology and returned frequent missives to Griffith containing his results. By another stroke of fortune, Griffith was appointed as one of the Commissioners of Railways in 1836 and persuaded his fellow members of the great importance of a geological map of the country as an aid to planning for the rapidly expanding railway network. Griffith's first map was published in 1838 at a scale of one inch to ten miles, and this was followed a year later by his splendid map executed at the scale of one inch to four miles. This map, and its various versions produced up to 1855, represents one of the milestones of Irish geology and remains one of the finest geological maps ever produced (Herries Davies, 1977a, 1983). While the map owes much to Griffith's own efforts, its genesis owes more to the geological observations made by Ganly (Archer, 1980), one of which deserves special mention. In 1856, Ganly published a short paper in the *Journal of the Geological Society of Dublin*, in which he illustrated the 'structure of strata' and demonstrated how the shape of rippled sandstone could be used to determine the original attitude of strata. His comments on what is now termed cross-stratification remain fundamental to geologists today (Wyse Jackson, 1995 [Chapter 11 herein]).

The body geologic

The university geologists, their colleagues in the Geological Survey, and amateur geologists such as Lord Talbot de Malahide, William Willoughby Cole (later the 3rd Earl of Enniskillen), and the Rev. Maxwell Henry Close, were brought together through their membership of the Geological Society of Dublin and its successor, the Royal Geological Society of Ireland. The Society, which was the third such body devoted to earth sciences in Britain and Ireland, had been founded in 1831, largely due to the efforts of Bartholomew Lloyd, the Provost of Trinity College, Dublin (Herries Davies, 1965). This Society acted as a forum in which geological opinions could be voiced and published in the pages of its *Journal* (Archer, 1987). It formed its own museum (M'Coy, 1841), which in 1844 was given to Trinity College in return for permission to hold its meetings there. By Royal Charter, it became the Royal Geological Society of Ireland and remained active until 1890.

The Society met once a month between November and June and as many as five papers were read at each meeting; nearly half were subsequently published. In the pages of the *Journal*, many important papers appeared: William Hellier Baily on fossil plants from the classic site at Kiltorcan, Co. Kilkenny (Herries Davies, 1990a); John Ball on the glaciation of Kerry (Herries Davies, 1990b); Maxwell Close on glacial striations near Dublin; Ganly on determining the way-up in strata; Griffith on the subdivision of the Carboniferous rocks in Ireland; Haughton on geological time; Hull on the microscopic nature of granites; Oldham on the Cambrian rocks at Bray; Kinahan on Irish economic geology; and Jukes on river valleys in southern Ireland. Oldham's and Kinahan's papers are of special significance. The work by Kinahan on the economic geology of Ireland was published in several papers which made up the eighteenth and final volume of the *Journal*. It contained information on the quarries of Ireland and described the rocks that were extracted from them. It remains a very important source for those interested in building materials. The paper by Jukes, read to the society on 14 May 1862, explains the curious nature of some rivers in Munster, most notably the River Blackwater, which flows due east until it reaches Cappoquin, where it turns through ninety degrees and flows due south to Youghal. The prevailing view at the time was that the landscape had been shaped by marine currents and waves during recent submergence. Jukes recognised the pivotal role of fluvial action in moulding present features and said that the drainage system adopted by these rivers was independent of the underlying geological structure. The Blackwater south of Cappoquin was a survivor of an old north-south system founded on the young southward-dipping surface, but above Cappoquin it was a younger river that extended itself westwards by headward extension along the older grain of the landscape (Herries Davies, 1985b).

Another influential member was Robert Mallet, an engineer who had a large ironworks in Dublin (Cox, 1982). Mallet, who served as President of the Geological Society of Dublin and its successor on three occasions, was interested in the mechanisms of earthquakes and conducted experiments on Killiney Beach to see how seismic waves moved through the sand. He compiled a seismic map of the world and was the first to measure the epicentre of an earthquake (Cox, 1985). He is today rightly regarded as the 'father of seismology' (Dean, 1991).

The British Association for the Advancement of Science was founded in York in 1831. Four years later, it met in Dublin under the presidency of Bartholomew Lloyd. This meeting generated a great deal of attention, much of which was beneficial to Irish geologists. Many of the eminent British geologists attended, including Sir Roderick Murchison, the Rev. Adam Sedgwick, and William Smith. Smith, who was a surveyor, had found lasting fame as the author of the first large-scale geological map of England and Wales, which he published in 1815. At the 1835 meeting, he was conferred with the degree of Doctor of Laws—his only academic accolade (Herries Davies, 1969a; Wyse Jackson, 2016 [Chapter 6 herein]). Since 1835, the Association has met in Ireland eleven times—in Belfast in 1852, 1874, 1902, 1952, and 1987; in Cork in 1843; in Dublin in 1857, 1878, 1908, 1957 (Morrell, 1990), and most recently in 2005—and has helped to generate enthusiasm among the native scientists.

After the 1835 meeting, Murchison and Sedgwick, together with Charles Lyell, John Phillips (Smith's nephew), Phillip de Malpas Grey Egerton, Louis Agassiz (an expert on fossil fishes), and Griffith, met at Florence Court, near Enniskillen, Co. Fermanagh, the home of William Willoughby Cole. Cole, together with Egerton, had accumulated one of the finest collections of fossil fishes on these islands, which was housed in its own museum in one of the pavilions at Florence Court (Herries Davies, 1970a). Such geological collecting did not impress Cole's father, who referred to it as being 'damned nonsense' (James, 1986). Many of the specimens were collected in the company of Egerton, and as the fish were often preserved as two halves on a split slab, it was customary for the two friends to keep one half each. Just before his death, Cole sold his collection to the British Museum (Natural History) (now the Natural History Museum) in London, where his specimens were reunited with those of Egerton who had earlier donated his collection.

University geology

The teaching of geology was placed on a firm footing in the 1840s when several chairs were founded in Irish universities. The oldest is in Trinity College, Dublin, where the chair was established in 1843; those in the Queen's Colleges followed in 1849. In Trinity College, geology had been taught from the 1770s onwards (Wyse Jackson, 1994a, b). Whitley Stokes lectured on natural history from the early 1800s and published two fine catalogues of minerals in the University Museum (Stokes, 1807, 1818). Although a classics scholar, William Fitton was interested in the geology of the Dublin area and as an undergraduate was arrested in the Dublin Mountains while engaged in fieldwork. His crime was possession of a dangerous weapon—his geological hammer! (Herries Davies, 1983; Wyse Jackson, 1998a [Chapter 5 herein]). Fitton, who produced a fine memoir on the geology of the region based on his, and the late Walter Stephens, observations, was an early member and President (1827–1828) of the Geological Society of London and through his writings in the *Edinburgh Review* raised the profile of geology in the Britain and Ireland (Stephens and Fitton, 1812; Woodward, 1907; Wyse Jackson, 1998a).

From 1777, Trinity College actively acquired geological specimens for its museum, which had been founded by William Hamilton (Wyse Jackson, 1992b) and which was used for teaching. The most important acquisitions were the minerals purchased in 1823 from the Right Honourable George Knox, one-time Member of Parliament for Dublin University (Wyse Jackson, 2006b [Chapter 21 herein]), and the Carboniferous fossils donated by Sir Richard Griffith in 1845. The Chair at Trinity was established in December 1843 and was offered to, and accepted by, John Phillips, Professor of Geology at King's College, London, and later at Oxford. Phillips thought he would be offered the additional post of Local Director of the Geological Survey, but in this he was disappointed. He resigned and returned to England—it was potentially a great loss for Ireland. He was succeeded by Thomas Oldham, who concurrently held the Geological Survey position. From 1851 until 1881, the Professor of Geology was the Rev. Samuel Haughton, a very able scientist whose interests were wide-ranging. He is remembered for his estimates of the age of the earth, which ranged from 2,298 million years to 200 million years, his work on animal mechanics, his opposition to Charles Darwin's views, and his invention of a humane method of execution by hanging (McMillan, 1988; Wyse Jackson, 1992b; Spearman, 1990).

In 1849, Chairs of Geology were established in the Queen's Colleges of Belfast, Cork, and Galway. Two able palaeontologists—Frederick M'Coy and William King—were appointed to Belfast and Galway, and a noted mineralogist, James Nicol, was appointed to the Cork chair. Nicol spent only four years in the position, during which time he published a manual of mineralogy, but he wrote little on the geology of Ireland and was succeeded by Robert Harkness in 1853. William King, who was born in north-east England (Pettigrew, 1979; Harper 1988a), is remembered as the author in 1850 of the seminal monograph *The Permian Fossils of England* (King, 1850), for his interest in coccoliths and as a principal contributor to the debate over Eozoon (DeArce and Wyse Jackson, 2014), which had been regarded as an early fossil but was shown by King to be composed of inorganic material (Harper, 1988b; Adelman, 2007), and for first naming *Homo neanderthalensis* (Murray *et al.* 2015).

M'Coy had a chequered career in Ireland; while still very young he was appointed Curator of the Geological Society of Dublin, the collections of which he catalogued (M'Coy, 1841). However, he fell out with the Society and was then employed by the Geological Survey for a short time before resigning in 1846 (Herries Davies, 1983). M'Coy then spent a number of years working in Cambridge, before moving to Belfast. He produced monographs on the Carboniferous (1844) and Silurian (1846) fossils of Ireland, based on material collected by Richard Griffith's valuation surveyors, that represent with little doubt the most important palaeontological publications ever produced in Ireland (Wyse Jackson and Monaghan, 1994 [Chapter 10 herein]). In general, the descriptions are short, but they are accompanied by accurate lithographs executed by Du Noyer. It is now acknowledged that in these works, more new species were described than can be justified by the material; nevertheless, they remain, together with the original specimens now in the National Museum of Ireland, important to present-day palaeontologists as sources of many valid fossil groups. M'Coy resigned his chair at Queen's in 1854 and took up the position of Professor of Natural Science in Melbourne, Australia, where he took on additional official positions (Vallance, 1978) and was knighted in 1891. However, after his departure from Belfast, his research output declined and the promise of his early career in Ireland was left largely unfulfilled.

Decline of Irish geology, 1890–1922

In 1890, the final map in the Geological Survey's mammoth series of one-inch-to-one-mile maps was published. Immediately, the Survey's staff was trimmed; the Dublin office was kept open with a skeleton staff who worked with decreased funding. Mapping of solid geology was no longer a priority; the general consensus was that everything was known of Ireland's geology. Nevertheless, during the period 1890–1922, the Geological Survey of Ireland carried out some important work, particularly in the area of glacial geology. Much of the impetus for this project came from George William Lamplugh, who took charge from 1901 to 1905. Lamplugh, who became Assistant Director of the Geological Survey of England and Wales, was a glacial geologist of considerable influence whose ideas were based largely on field observation (Penny, 1966). Together with men such as William Bourke Wright, James Robinson Kilroe, Herbert Brantwood Muff, Henry J. Seymour, Sidney Berdoe Neal Wilkinson, and Alexander M'Henry, Lamplugh surveyed the glacial deposits, revised the solid geology of the major metropolitan areas, and produced maps and memoirs of Dublin (1903), Belfast (1904), Cork (1905), Limerick (1907), and Londonderry (1908).

After 1905, the Irish section of the Geological Survey became an autonomous entity under the directorship of Grenville Arthur James Cole (Wyse Jackson, 1989, 2007b). During the next twenty years, the Geological Survey of Ireland produced a number of *Memoirs* on economic topics: the intrabasaltic rocks of Antrim (1912), the geology of the Ballycastle Coalfield (1924), and the mines and mineral deposits of Ireland (1922). Solid geology was not neglected; Clare Island (1914) and the Killarney region (1927) were surveyed. Possibly the most important publication of this period described rocks from the seabed of the Porcupine basin off the west coast of Ireland, representing an early examination of the geology of an area that recently has been the focus of considerable exploration in the search for hydrocarbons (Cole and Crook, 1910).

The slow demise of the Geological Survey coincided with a general decline in Irish geological fortunes. The Royal Geological Society of Ireland had for a number of years prior to 1890 been affected by falling membership, resulting in financial difficulties which caused delays in publishing parts of the *Journal*. This in turn probably resulted in further non-renewed memberships. Finally, in 1894, the Society ceased to exist, having stopped meeting four years earlier. It was another sixty-five years before another such body donned its tattered mantle. It is to the credit of both the Royal Irish Academy and the Royal Dublin Society that they continued to provide a channel through which Irish geologists could publish their research findings. During this period, the monthly journal *The Irish Naturalist* appeared. This acted as a catalyst for naturalists, particularly those associated with the Field Clubs, and enabled them to publish their research findings (Wyse Jackson and Wyse Jackson, 1992). However, it was a 'popular' journal and precious little of geological note was published, other than a number of papers by Grenville Cole.

In some of the universities, the appointment of academics to geological positions was not, in hindsight, beneficial. Excellent researchers such as Samuel Haughton, William Johnston Sollas, an expert on fossil sponges and a keen field geologist in Trinity College, and Andrew Leith Adams, a medic, palaeontologist, and cave explorer in Queen's College, Cork, were replaced by men such as John Joly in Trinity and Marcus Hartog in Cork. Although Joly was undoubtedly one of Ireland's greatest scientists, he was largely interested in physics and radioactivity (Dixon, 1934 1941; Nudds, 1986). Field geology did not enthuse him although it inspired him to write poetry on the early fossil *Oldhamia* (see Nudds, 1983; Pemberton, 2010; Wyse Jackson 2011 [Chapter 16 herein]), which is probably why few geologists of note graduated from Trinity College during his tenure of the Chair of Geology and Mineralogy from 1897 to 1933 (Wyse Jackson, 1992a). Joly is perhaps best remembered for his pioneering work in colour photography (Coonan, 1991) and for developing a method of treating cancer with radium. His geological work included research into the age of the earth (Wyse Jackson 2001a [Chapter 17 herein]) and the relationship between radioactivity and geology (Nudds, 1986; Wyse Jackson 2002a).

Hartog was Professor of Natural History at Queen's College, Cork (which became University College in 1909) from 1882 to 1909 and then became Professor of Zoology until 1921, but in that time he published only one geological paper. The major university geologist of note in Ireland at this time was Grenville Cole, who was Professor of Geology at the Royal College of Science for Ireland in Dublin as well as Director of the Geological Survey of Ireland (Wyse Jackson, 1989). Cole was small in stature but had unlimited energy. He was a prolific author, with over 500 publications to his name, and an advocate of field studies, which he stressed were of fundamental importance for an understanding of geology. He was also a keen tricyclist and later bicyclist who together with his wife made frequent excursions deep into Europe (Wyse Jackson, 1991, 2007b [Chapter 18 herein]). He played an important role in the development of geography in Ireland, urging its adoption as a university subject (Herries Davies, 1977b).

In 1902 and 1908, the British Association for the Advancement of Science visited Belfast and Dublin, respectively, and while these meetings attracted many scientists of note and undoubtedly raised the interest of those in Ireland, they failed to arrest the slow erosion of interest in geology in Ireland. There were at this time a number of amateur geologists who made important contributions to Irish geology. The majority hailed from the Belfast area, where they had come under the influence of Ralph Tate and later the Belfast Naturalists' Field Club. Two women, Sydney Mary Thompson (Madame Christen) and Mary Andrews, collected information on glacial erratics which was presented to the British Association (Higgs and Wyse Jackson, 2007; Wyse Jackson 2009a [Chapter 19 herein], 2019b). Joseph Wright was a Quaker grocer of Belfast who became an expert on microscopic foraminifera, particularly those recovered from Cretaceous strata in north-east Ireland, and from glacial boulder clays as far afield as Canada and the Arctic (Wilson, 1987). Robert John Welch, another amateur naturalist, was the foremost photographer of the period; his photographs of geological phenomena plants, scenery, ships and shipbuilding, and prehistoric monuments appeared in many important publications as well as popular handbooks produced by railway companies (Evans and Turner, 1977). He spared no effort in getting the best image and would spend considerable time setting up his heavy equipment before shooting. He developed and printed all his own photographs, the clarity of which would defy many modern photographers. Welch's photographs were all the better for a keen perception of his subjects. He made the study of molluscs his prime natural history pursuit and was awarded an honorary MSc by Queen's University (Praeger, 1949). Over 5,000 of his photographic plates are now in the Ulster Museum, as are the many photographs of geological interest that he presented to the British Association (Turner, 1979; Hackney et al., 1983; McKenna, 1990). The Irish prints in the Ulster Museum number 638, of which Welch contributed over half. Others were taken by Grenville Cole, William Gray (a Belfast naturalist), Mary Andrews, and Professor Sidney Hugh Reynolds.

Reynolds, who was Professor of Geology at Bristol University, was one of the few England-based geologists who worked in Ireland around the turn of the century. Together with Charles Irving Gardiner, he comprehensively mapped various Ordovician and Silurian rocks in Ireland and published nine papers reporting their findings, which were important at a time when the relationship between rocks of the two geological periods was first being understood. After 1922, Reynolds and Gardiner ceased their Irish work and turned their attention to matters in Britain (Wyse Jackson, 2009b).

Another blow to the geological community came in 1922 when the new Irish Free State government decided to take over the geological galleries of the Museum of Science and Art for use by clerical workers. The scientifically and historically important collections of Irish and foreign geology, acquired over nearly two centuries, were put into storage where they remained for many decades, and only today are they beginning to be seen again.

Why did geology decline from the 1890s? There are a number of reasons: the completion of the Geological Survey mapping programme; the constant emigration of competent geologists and geology graduates, which increased after 1922 as many Protestants left Ireland; and the effects of the First World War, when at least three budding Irish geologists were killed along with thousands of other young men on the fields of France.

Irish geology in the doldrums 1922–1960

While geology in official government institutions was not regarded as of national importance, the universities continued to train geologists. Many of these graduates of Trinity College, Dublin, Queen's University, Belfast, and the University Colleges of Cork, Dublin, and Galway emigrated soon after graduation and found geological work in various parts of the globe. Few stayed at home because there were no employment opportunities, either economic or academic, in Ireland. Between the 1920s and the 1950s, university geology was dominated by a handful of academics: Louis Bouvier Smyth at Trinity College, John Kaye Charlesworth at Queen's University, Belfast, Henry B. Seymour at University College Dublin, and Isaac Swain at University College Cork, each of whom largely operated as one-man departments. With the exception of Charlesworth, these leading figures of Irish geology did not publish a great deal and consequently were not well known beyond Ireland's shores.

Charlesworth, a Yorkshireman, was appointed as Professor of Geology at Queen's University, Belfast, in 1921 and thus became the first to hold the Chair since M'Coy's departure in 1854. He wrote extensively on the Pleistocene of Ireland and other areas, and his work culminated in 1957 in the bulky two-volume *The Quaternary Era* (Wilson, 1972; Williams, 1973). Smyth, who was a quiet, unassuming man, is remembered for his work on coral faunas from the Lower Carboniferous rocks and for a fine and meticulous stratigraphical description of Hook Head, Co. Wexford (Farrington, 1954). Swain, a Quaker, published only a handful of papers during his tenure in Cork as Professor of Geology and Geography from 1908 to 1944.

In the 1940s, some of the university departments were enlarged through the employment of further academic staff. At the same time, the first postgraduate students were enrolling in the universities. This period marks the slow climb of Irish geology out of the trough in which, it had languished during the period between the two world wars. Through the 1950s, geological graduates still found most employment opportunities abroad, although more stayed in Ireland—initially, at least, to carry out postgraduate research. It was during this period that William Edward Neville of Trinity College was awarded the first PhD in Irish geology in 1951. From 1951 to 1992, over 200 higher degrees have been awarded in geology on the basis of research theses: ninety-four from Trinity College; sixty-three from Queen's University, Belfast (since 1969); twenty-eight from University College Dublin; twenty-two from University College Galway; and twenty from University College Cork. Many more have been awarded since. These postgraduate theses have added considerably to our present knowledge of Irish geology.

The renaissance of Irish geology from 1960

Since the early 1960s, the state of Irish geology has improved enormously. There are a number of reasons for this: increased staffing in universities, discoveries of major base metal and hydrocarbon deposits in Ireland, confidence in the merits of exploration, and the development of professional and semi-professional associations to cater to the expanded geological community.

Without exception, staffing in the five university geological departments has increased, the research output has diversified, and, in consequence, the number of publications has jumped. Such research has been of immense value in increasing the sphere of knowledge pertaining to Ireland's geological structure and has been published in international and Irish earth science journals such as the *Scientific Proceedings of the Royal Dublin Society*, the *Proceedings of the Royal Irish Academy*, and the *Irish Journal of Earth Sciences*. The last of these journals was established in 1978, published by the Royal Dublin Society until 1983 and by the Royal Irish Academy since then. It maintains a link between geologists working in Ireland today and some of the pioneers of nearly 200 years ago who presented geological findings to these two bodies.

The second factor in the revitalisation of Irish geology was the realisation and subsequent confirmation that Ireland possessed significant deposits of economically viable resources, both onshore and offshore. In 1960, the government of the Republic of Ireland passed the Petroleum and Other Minerals Development Bill, which allowed for the exploration of parts of Ireland and its surrounding continental shelf. New techniques, such as gravity measurements and seismic profiling (the method first used by Robert Mallet on Killiney Beach nearly 100 years before), were employed for geological exploration. In 1961, the first significant find—large reserves of base metals at Tynagh, Co. Galway—was announced. Exploration by various mining companies continued throughout the 1960s and further base metal deposits were discovered at Navan, Co. Meath, and Gortdrum, Co. Tipperary, which like Tynagh were brought into production, as were deposits at Avoca, Co. Wicklow, and Silvermines, Co. Tipperary. More recently, base metal deposits have been discovered at Lisheen, Co. Tipperary, and Galmoy, Co. Kilkenny.

The search for oil and gas in Ireland from the early 1960s was prompted by the discovery of gas in the Netherlands. Initially, exploration and drilling were carried out onshore in Leitrim and Fermanagh, but soon interest was focused on the continental shelf. Such exploration was controlled at the government level by the provisions of the Continental Shelf Act of 1968, and in 1971 a large gas field was discovered in the north Celtic Sea Basin, twenty-five miles south-west of Kinsale, which for many years supplied much of the Republic of Ireland's domestic demand for natural gas.

The increasing importance of economic geology was reflected in the foundation of the Irish Association for Economic Geology (IAEG) in 1974, which aimed to advance the practice of economic geology, provide information regarding mineral and hydrocarbon exploration and production, and promote the exchange of ideas and information through field excursions and meetings. IAEG produces an annual review and has published two fine volumes, *Geology and Genesis of Mineral Deposits in Ireland* (1986) and *The Irish Minerals Industry—A Review of the Decade* (1992).

While the credit for the discovery of many of the hydrocarbon and metal deposits must go in part to the exploration companies, the university geological departments and various government agencies played an important role as well. The Geological Survey of Ireland has embarked on compiling a new series of geological maps of the country, at the scale of 1:100,000, that will complement maps recently produced by the Geological Survey of Northern Ireland.

The last factor in the renaissance of Irish geology was the foundation of the Irish Geological Association—the modern-day successor of the Royal Geological Society of Ireland. This was established with the encouragement of Robert George Spencer Hudson, who was Professor of Geology at Trinity College, Dublin, from 1961 until his death in 1965. He had been Professor of Geology at Leeds and later worked in the Middle East before coming to Dublin. Hudson became an authority on Carboniferous stratigraphy and coral faunas. He fostered in his students a love of research and field geology (Jackson, 1966). The Association, which brings together professional and amateur geologists from all parts of the island, organises lectures and field excursions, an annual research meeting at which new findings are presented, and an annual geology day at which the general public is invited to participate in a wide range of geological activities such as field trips and museum visits.

Despite its small size, Ireland has made important contributions to the study of glaciation and glacial geology. Early writings included those of Richard Prior, who described an esker in 1699, and Charles Smith, who was the first to describe a drumlin swarm as a 'basket of eggs' topography in 1744 (Herries Davies, 1966, 1970b). Indeed the words 'drumlin' and 'esker' are derived from the Irish language—druim meaning back and eiscir meaning a ridge (Quin and Freeman, 1947). As we have seen, John Ball and Maxwell Close in the nineteenth century made important observations on glaciation, and Ireland was well known throughout Europe for the Giant Irish Deer *Megaloceros giganteus*, which had become extinct at the end of the last glaciation 10,000 years ago, and of which many skeletons were found beneath a raised bog at Ballybetagh, Co. Dublin. But the study of glacial geology only developed a measurable popularity following the efforts of the Royal Irish Academy, which brought Knud Jessen to Ireland from Denmark in the 1930s. After these visits, various native students of the Irish

Pleistocene emerged, most notably Anthony Farrington, a graduate of University College Cork and subsequently Resident (later Executive) Secretary of the Royal Irish Academy (Mitchell, 1990), Frank Mitchell and William A. Watts of Trinity College, Dublin, and Francis Millington Synge of the Geological Survey of Ireland, who together made major advances in our knowledge of the effects of the last glaciation in Ireland. Today, glacial geologists and botanists have their own organisation, the Irish Association for Quaternary Studies.

In the last forty years, geology in Ireland has been thriving. Great advances in our knowledge of the island's geological structure have been made. History will relate which aspects of research made in this time will have had the greatest effect, but perhaps among them may rank the following: the research carried out on the Donegal granite by Wallace Pitcher and others from Liverpool; the research by Bernard Leake and colleagues from Glasgow on the complex geology of Connemara; the collection and analysis of the vast volume in geological data derived from offshore Ireland and from the exploration for metals onshore; research on Lower Palaeozoic rocks and fossils which has yielded much information on the palaeogeography of Ireland; the work on thermal maturation of the Upper Palaeozoic rocks by researchers in Trinity College, Dublin; advances made in structural geology which have contributed greatly to understanding Ireland's underlying geology; the work carried out by Mitchell and others which has opened our eyes to the complexity of the last Ice Age and the greening of Ireland after the ice had melted; and finally the work *A Geology of Ireland* (Holland, 1981), and its two successive editions (Holland, 2001; Holland and Sanders, 2009) syntheses of much of this recent knowledge into a comprehensive and usable book.

In 1856, after a three-week visit to Ireland during which he suffered from the enveloping mists of the Dingle peninsula, the Director of the Geological Survey, Sir Roderick Murchison, returned to England and declared

> that the geology of Ireland is the dullest which I am acquainted with in Europe. If St Patrick excluded venomous animals he ought to have worked a miracle in giving to the holy isle some one good thing under ground.

A great deal has happened in Irish geology since.

1700s–1800s

2

TUMULTUOUS TIMES: GEOLOGY IN IRELAND AND THE DEBATE ON THE NATURE OF BASALT AND OTHER ROCKS OF NORTH-EAST IRELAND BETWEEN 1740 AND 1816

In P.N. Wyse Jackson (ed.) *Science and Engineering in Ireland in 1798: a time of revolution.* (2000). Royal Irish Academy: Dublin, pp. 35–49.

Introduction

This paper focuses on the debate on the nature of rocks on the north-east coast of Ireland in the late eighteenth and early nineteenth centuries and examines the role of protagonists on both sides, especially that played by the Reverend William Richardson (1740–1820).[1]

The debate on the nature of basalt in north-east Ireland

The Giant's Causeway and surrounding area had attracted the members of the short-lived Dublin Philosophical Society (1683–1708), some of whom attempted descriptions of its features—namely Sir Richard Bulkeley in 1693, Bishop Samuel Foley (1694), and Thomas Molyneaux (1694) Later, the Lord Bishop, William Augustus Hervey (1730–1803), Bishop of Derry and fourth Earl of Bristol, apparently installed steps down to the Giant's Causeway and inspired made trips to the volcanic regions of Italy (Childe-Pemberton, 1925).

[1] When originally published the paper included a long discussion of the state of geology in Ireland in the 1790s and commented on geological activity in the Dublin Society, Trinity College Dublin, and the Royal Irish Academy. These sections are omitted here as they are adequately covered in Chapter 1 and in others in this volume.

 DOI: 10.4324/9781003475552-5

Figure 2.1 East Prospect of the Giant's Causeway. Engraving 1777 by Vivares after Susanna Drury (Author's collection).

It was, however, the award of a premium to Susanna Drury, a Dublin artist, that accelerated interest in the area. She spent a period of three months painting various views of the Giant's Causeway providing excellent representations of the coastline and of the pillars of columnar basalt (Anglesea and Preston, 1980). Subsequently, her images were engraved in London by Francois Vivares (Figure 2.1), and viewing of his prints and others initiated a steady stream of visitors to the Causeway Coast, whose rocks then became the subject of examination and explanation. Visitors included Archdeacon Richard Pococke in 1747. He suggested that vertical pillars (of basalt) formed as precipitated muds became compressed by overburden pressure into the polygonal shapes typically seen at the Giant's Causeway (Pococke, 1748a, b, 1753), John Wesley in 1778, John Whitehurst in 1783, Abraham Mills in 1787–1788 (Herries Davies, 1978a), the Frenchman, De Latocnaye in 1797, Humphry Davy in 1806 who may have been guided by William Richardson (Siegfried and Dott, 1980) (Figure 2.2), and the Swiss geologist Jean-Francois Berger in 1811 (MacArthur, 1990).

In Ireland, the debate on the origin of basalt and other rocks naturally centred theoretically and petrologically on whether these rocks were sedimentary or igneous in origin, and geographically around the Giant's Causeway. One body of thought, the 'Neptunists' or to use a contemporary term 'watermen', felt that these rocks were precipitates from water—a view held by the German mineralogist Abraham Gottlob Werner (1749–1817) and his students in the Freiberg School of Mines. They regarded lava as being a product of volcanoes but held that where volcanoes occurred in basaltic regions, they had formed after the basalt had been deposited. Some were convinced that lithologically basalt was quite distinct from lava and others that lava was produced when volcanic heat and air acted on pre-existing (presumably water-lain) basalt. The other group,

Figure 2.2 View of Benmore (Fair Head) on the north Antrim coast, showing moderately well-developed columnar basalt. It is possible that the figures on the rocks in the foreground are the Rev. William Richardson (pointing out geological structure) and Humphrey Davy (seated and sketching). Original drawing by D. Stewart (possibly the Itinerant Mineralogist employed by the Dublin Society). Engraving reproduced from William Hamilton Drummond, 1811. *The Giants' Causeway, a poem*. Belfast: Longman, Hurst, Rees, Orme & Browne (Author's collection).

variably known as 'Vulcanists' or 'Plutonists' or 'firemen', according to the rock types in which they were interested, argued that the rocks were once heated and were the products of volcanoes or other igneous mechanisms (Herries Davies, 1981). The Vulcanist theory was proposed by the Frenchman Nicolas Desmarest (1725–1815), who, having observed the extinct volcanoes in central France, advocated in 1768 that basalt was the product of volcanic activity (Taylor, 1969, 2007). Taylor points out that Desmarest's first public airing of his views came in 1765 but was published three years later in the *Encyclopedie* (Desmarest, 1768) and received widespread attention with the publication of his subsequent monograph on the subject (Desmarest, 1771). Desmarest's conclusions on the origin of basalt were quickly affirmed by others including Sir William Hamilton (1730–1803) and John Strange (1732–1799) who were familiar with the volcanic regions in Italy. In 1788, James Hutton (1726–1797), the Scottish geologist, published his paper *Theory of the Earth* which proposed, amongst other things, that some igneous rocks were plutonic in origin and argued that these were emplaced from a hot magmatic source from below, and that large-scale reworking of older rock successions had taken place in the past. The source of the heat was a matter of further debate until the late eighteenth century. Some argued that the source of heat that produced the igneous rocks was derived from a deep source while others suggested that the source was shallow from combustion of coal and oil (den Tex, 1996).

Firemen vs Watermen in the Irish context

Nicolas Desmarest never viewed the Giant's Causeway, but on seeing Vivares' engravings of Susanna Drury's paintings was the first geologist to declare that the basalts that formed the region were volcanic in origin. John Whitehurst (1786) suggested that the columnar basalts of the Giant's Causeway formed as the molten basalt cooled. In the same year the Rev. William Hamilton, who had visited the region, published an important and influential memoir entitled *Letters concerning the Northern Coast of the County of Antrim* in 1786 which advanced Desmarest's theory of the volcanic origin of the Causeway rocks (Hamilton, 1786; Figure 2.3). Others, including John Wesley, regarded the basaltic successions in north-east Ireland as being the product of some volcanic fire.

Perhaps, the most influential member of the Dublin geological fraternity at the time was Richard Kirwan, President of the Royal Irish Academy from 1799 until 1812 (Scott, 1973, 1979; Wyse Jackson 1998c [Chapter 4 herein]; Newcomb, 2012). Although many of his ideas are now known to be incorrect, his ability to argue through the printed page forced others to look at these ideas and

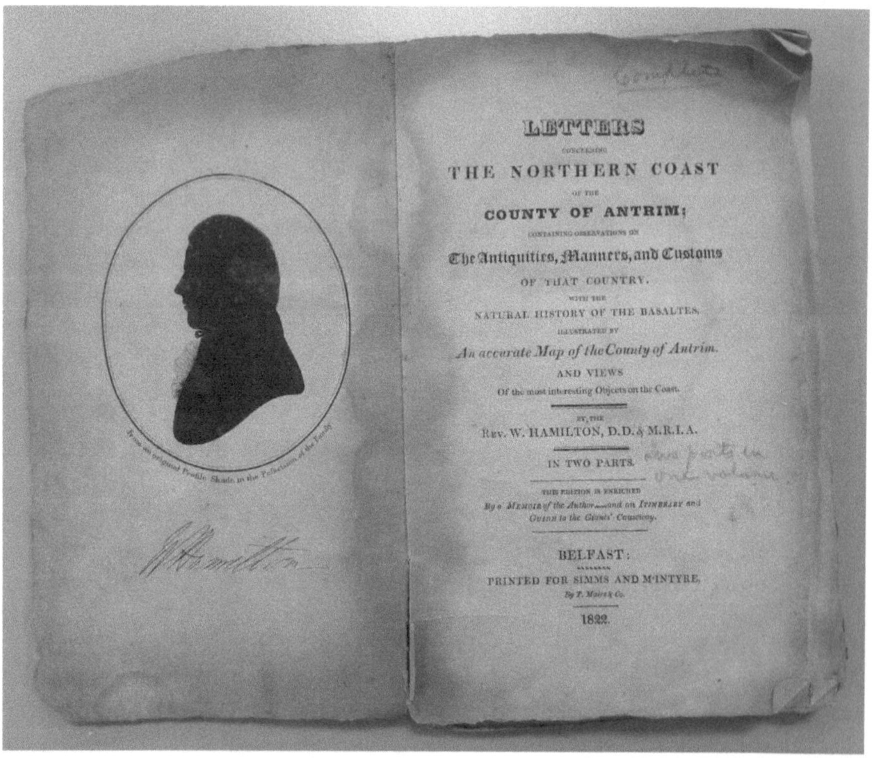

Figure 2.3 Rev. William Hamilton represented in an 1822 edition of his *Letters concerning the Northern Coast of the County of Antrim* (Author's collection).

counterargue. He was outraged at Hutton's ideas and went into print defending the Neptunian theory in a number of papers from 1793 and in his *Geological Essays*, published in 1799. Some of Kirwan's objections, which are somewhat difficult to glean from the contorted logic and language of his papers, can be summarised briefly: he said that basalt rested on various rock types, including coal, limestone, gneiss, and granite, in which no evidence of the effect of heat could be seen. It is now known that gneiss is one of the highest grade metamorphic rocks, which is produced by the effect of enormous stress and high temperature. He noted that many basalts contain zeolites, minerals rich in water and calcite, which on account of their chemistry could not have been heated. He failed to realise that the zeolites were deposited subsequent to the formation of the basalt. He also objected to the findings of Sir James Hall (1761–1832) who had taken basalt, melted it, and recrystallised it, in order to show the similarity of the artificial with the real (Hall, 1805, whose experiments were analysed by both Eyles, 1961, Wyllie, 1998). Here was good proof that basalt was igneous in origin. The recrystallised basalt contained minute air bubbles which Kirwan naturally stated were characteristic of basalt deposited in water.

While Kirwan was loud in his rejection of Hutton and Desmarest, his disciple the Reverend William Richardson, was louder, and perhaps with good reason, for he actually spent some of the year living at Portrush, on the Antrim coast. Richardson, a Fellow of Trinity College of 1766, took Holy Orders and eventually resigned in September 1783 on taking up the living of the parish of Clonfeacle in County Tyrone. Through his membership of the Royal Irish Academy, he would have become familiar with the controversy surrounding the Giant's Causeway. During his annual sojourns at Portrush, Richardson examined the ground for himself, and he published a steady stream of papers between 1802 and 1812 which outlined his views on the geology of the region (Richardson, 1802, 1803a, b, 1805, 1806a, b, 1808, 1812). He clearly was upset at others pontificating on the geology of an area which they had either not visited at all or had examined for just a short time. He believed he had the authority, as a long-term resident, to make his views known. While he claimed not to subscribe to any theory or belong to any faction (Richardson, note 43 in Dubourdieu, 1812), he clearly if unwittingly held to the Neptunist theory for the origin of the basalt. He rejected the alternative theories on a number of grounds:

1. There was no evidence of a volcanic mountain or cone in Antrim.
2. Plants are found developed on lava flows but not in between the basalt layers.
3. The constituents of basalt and lava were different.
4. The layers of basalt were horizontal and regular in thickness.
5. The physical appearance of basalt and lava were different.
6. The contact relationships of basalt and lava were different.
7. The basalt was divided into regular masses while lava is found as an irregular mass.
8. The basalt of Portrush contained fossil Marne shells.

Given the limited exposure, Richardson had to other volcanic regions his objections to the Vulcanist and Plutonist theories are not surprising. The ammonite-bearing lithology at Portrush was discovered by Richardson in about 1799 and was proof to him and many others of the sedimentary origin of basalt. He also located specimens of basalt in a quarry at Ballylagan in which cavities were infilled with 'fresh water'. To him this was stronger evidence of the sedimentary origin of basalt than was the presence of fossils. Richardson's material was dispatched to Kirwan and other geologists, and later to Plutonists in Edinburgh. He also sent twenty-two specimens of basalts and ammonite-bearing rock to Trinity College Dublin where it was catalogued by Robert Perceval; the water-bearing basalt was number four. Unfortunately, this suite of specimens cannot currently be recognised in the Geological Museum, although Perceval's catalogue survives.

Aside from his geological opinions, Richardson involved himself in politics (Blackstock, 2013) and turned his pen to agricultural subjects and wrote extensively on the merits of growing fiorin grass (see Nelson, 2000 for caricature of him scything grass). He died in September 1820, continuing to be the last to hold onto his outdated geological viewpoint.

Resolution and conclusions

Specimens of the fossil-bearing 'basalt' were sent to Edinburgh in 1801, where they were examined by Sir James Hall, Lord Webb Seymour, and John Playfair who identified the specimens as being sedimentary cherty hornstone (Playfair, 1802). The rock is now known to be Lower Jurassic Lias mudstone which contains specimens of the ammonite *Paltechioceras*. The Portrush rock had been baked following intrusion of the dolerite Portrush Sill and this resulted in it taking on a similar appearance to that of basalt. Their conclusions were confirmed by Jean-Francois Berger and by the Reverends William Conybeare (1787–1857) and William Buckland (1784–1856) in 1816. This piece of evidence was critical and swung the scales in the favour of the Vulcanists and Plutonists, whose view held that basalt was derived from a molten source.

Additional evidence on the true nature and origin of basalt came from later observations made in Italy and elsewhere of active and extinct volcanoes where the presence of columnar basalt was seen in unequivocal lava flows. In terms of the Antrim basalts, it is now known that they were vented from fissures and that a volcanic cone was never present in north-east Ireland; that plants are preserved in the interbasaltic horizons; and that the regularity of the columns is due to slow-cooling of a ponded viscous lava in a subsided sycline (Simms, 2021).

Today, the Giant's Causeway and Causeway Coast continue to attract and excite geologists and the general public alike and its importance was confirmed by its designation some twenty years ago as a UNESCO World Heritage Site and inclusion in 2022 as one of the first 100 IUGS Heritage Sites (Lemon *et al.*,

2022). While there had to be a winner and a loser in the debate on its origins and composition, the Causeway Coast in north-east County Antrim remains one of the most important sites in the history of geology. It is a matter of some pride, to Irish historians of science and geology, that much of the sabre-rattling pertaining to the debate on the nature of basalt during the late 1700s and early 1800s should have taken place partially in Ireland.

3

THE FOSSIL FISHES OF BOLCA AND THE TRAVELS IN ITALY OF THE IRISH CLERIC GEORGE GRAYDON IN 1791

Museologia Scientifica 12(1–2) (1995), 57–81 [co-authored with Ezio Vaccari].

Introduction

The scientific travels in Italy of George Graydon are an interesting and very little-known example of the growing interest in Italian geological features, which increased significantly during the second half of the eighteenth century within the Anglo-Irish scientific community. The building up of a specialised lithological and paleontological collection, such as the one which was brought to Ireland by Graydon, not only contributed to the spread of the geological knowledge of Italy but also supported the circulation of scientific theories linked to particular specimens. This paper, carried out with research in Ireland and Italy, provides information about the history of the diffusion of the fossil fishes of Bolca in Ireland since the end of the eighteenth century. Within the lively context of Irish science in these years, Graydon's description of the fossil fishes and his interpretation of their formation, together with the account of some Italian studies on the fossiliferous site of Bolca, represent an important confirmation of the high level of scientific communication established in late eighteenth-century Europe.

The life of the cleric George Graydon

Graydon was probably born in County Kildare, Ireland in about the year 1753, into a prominent landowning family. His parents were George Generosus Graydon (d. 1771) and Jane Graydon (née Maxwell), who had four sons and a daughter, and who lived at Killashee House (now largely rebuilt and occupied by a school), south-east of the market town of Naas. At the age of 16, Graydon entered Trinity College, Dublin in November 1769; he graduated in 1774 with a B.A. and took the degree of LL.B. in 1782 (Todd, 1869, p. 233). After his graduation, he took Holy Orders and was ordained a minister of the established church, the Anglican

DOI: 10.4324/9781003475552-6

Church of Ireland. From 1791 Graydon was successively Prebendary of two parishes situated in the centre of the medieval part of Dublin: St. Michael and All Angels (1791–1797), a small parish of just over 2,000 persons, and St. John's Fishamble Street (1797–1803) a larger parish twice the population of the former (Cotton, 1848, pp. 69, 80). He held these positions simultaneously with that in Burrishoole, Co. Mayo, a parish in the west of Ireland of which he was Vicar from 1793 at the latest until his death in 1803. He divided his time almost equally between Dublin and Mayo and always spent the winter months in the city, except for those years when he was abroad.

The parish of Burrishoole was situated in an area of poor agricultural land, where a large population lived in often squalid conditions on small holdings and survived on subsistence farming and fishing. Graydon, together with his wife Elizabeth (nee Knott), whom he married in 1795, attempted to alleviate the suffering and poverty of these people. He was concerned with improving fishing methods, and she established a straw bonnet-making industry in 1797 which employed 200 girls, who could earn up to 1 shilling a day (Graydon, 1802). The efforts of the Graydons and particularly those of Elizabeth were rewarded by the award of a medal by the Dublin Society in November 1800 (Wyse Jackson and Vaccari, 1993, p. 94). In early 1803, George Graydon died, and he was probably buried at Gilltown, Co. Kildare, in the same graveyard as his parents. His widow was unfortunately left in straitened circumstances, as her husband had assumed the responsibility for debts incurred by his brother. She sold all his effects in order to raise money, but his estate showed a deficit of nearly £200, a considerable sum at the time. After 1804, all traces of Elizabeth Graydon have been lost.

Graydon's travels in Italy in 1791

When Graydon completed his studies in Trinity College, Dublin in the 1780s was emerging as a city of intellectual and scientific promise (Herries Davies, 1985c, pp. 297–302). The Dublin Society was founded in 1731 to promote science and agriculture, while the Royal Irish Academy emerged in 1785 for the promotion of the arts, polite literature, and science (Meenan and Clarke, 1981; O'Raifeartaigh, 1985). On the 14th of February 1786, Graydon was elected the 100th member of the Royal Irish Academy. Rapidly, he became active in the activities of the Academy, regularly attending its monthly meeting: on March 1789, he was admitted to the Council as a member of the Committee of Antiquities and early in 1790, he temporarily replaced the librarian of the Academy. A few months later, on the 26 June 1790, Graydon achieved the prestigious position of Secretary for Foreign Correspondence, an office he kept until the 16 March 1803, with only an interruption of one year between March 1791 and March 1792 (R.I.A. General Meetings Minutes, pp. 17, 45, 61, 67, 78, 187; R.I.A. Council Minutes, vol. 1, pp. 134, 169, 182).

Among the most distinguished members of the Royal Irish Academy were Richard Kirwan (1733–1812), the Rev. William Richardson (1740–1820), and the Rev. William Hamilton (1755–1797), all of whom were interested in the

Earth Sciences. It is in fact quite significant that in the first 11 volumes of the *Transactions of the Royal Irish Academy*, from 1787 to 1810, 45% of the papers in science were devoted to disciplines such as astronomy, geology, and meteorology (Herries Davies, 1985a, p. 247). This percentage seems to be even more remarkable when compared with the quantity of papers concerning Earth Sciences (about 7% of the total) published by the Royal Society of London between 1781 and 1820 (Torrens, 1994, p. 53).

The Irish scientists were particularly involved in the late eighteenth-century European debate between 'Neptunists' and 'Vulcanists' (or 'Plutonists', according to the rock types in which they were interested) on the nature and origin of basalt and other igneous rocks. Kirwan and Richardson followed the Neptunist view, held by the German mineralogist Abraham Gottlob Werner (1749–1817), who felt that basalt and the majority of rocks were precipitated from water. On the other hand, Hamilton, who wrote a fine and important memoir on the north coast of Antrim (Hamilton, 1786), accepted the Vulcanist theory, which argued that basalt and other rocks were once heated and were the products of volcanoes and other igneous phenomena (Herries Davies, 1978a, pp. 88–91, 1985a, pp. 252–257).

It was the geology of the north coast of Antrim in Ireland, with its famed basaltic outcrop at the Giant's Causeway, that became important for resolving the debate which eventually came down on the side of the Vulcanists. The contribution made by Irish scientists was widely known in Europe through the *Transactions of the Royal Irish Academy* in which various protagonists of the controversy wrote, particularly Kirwan, who became highly regarded (Herries Davies 1978b, pp. 13–14; Wyse Jackson, 1998c (Chapter 4 herein)).

George Graydon was certainly influenced and stimulated by the geological discussions on volcanic rocks which took place in the meetings of the Academy during the late 1780s, but his idea of travelling to Italy was probably also linked to the attempt to realise an academic project. In April 1790, the Council of the Academy decided to establish a general museum for the use of their members, but prior to going ahead with this plan they sought the attitude of the Dublin Society which already had a museum, regarding a possible clash of interests (R.I.A. Council Minutes, vol. 1, p. 179). Graydon was asked to get an estimate for fitting out two rooms as a museum of Irish antiquities, geology and natural history, and he also presented a plan for obtaining a cabinet of mineralogical specimens, which was considered by Richard Kirwan and accepted by the Council. A cabinet was manufactured by Mr Thorndale for the sum of 23 guineas and was in place in the Museum in November 1790 (R.I.A. Council Minutes, vol. 1, pp. 184, 190). However, it contained no specimens.

This probably prompted Graydon to set about procuring specimens himself for the Academy. He planned his trip to Italy to observe at first hand the features of the Italian volcanic terrains and to collect rock samples from Vesuvius and the area of the extinct volcanoes in the Venetian Republic, north-east of Italy. He did this in order, as has been seen, not only to fill the cabinet in the Academy's museum, but also to make available information to its members who could not travel to the region for themselves. He also wished to understand the origins and nature of Vesuvius and extinct volcanoes and to correlate and draw any analogies with the basaltic region in north-east Ireland.

The information traced to date about the period of this trip is still fragmentary, but we can assume that Graydon left for Italy in late 1790, as he himself stated that he was already in Verona in January 1791 (Graydon, 1794, p. 282). A few months later, on 16 March 1791, because of his prolonged absence from Dublin, the Royal Irish Academy temporarily removed him from the office of Secretary for Foreign Correspondence and nominated Robert Perceval, a chemist, in his place.

The Italian travels of George Graydon are documented in his two printed papers (Graydon, 1794, 1816) and in other handwritten material, such as a letter from Graydon to the Derbyshire geologist White Watson (1760–1835), a diary of the visit on Vesuvius and a catalogue of the collected specimens (Wyse Jackson and Vaccari, 1993, pp. 91–93). From these sources, it appears that the Irish cleric stayed for a short time in the north-east of Italy early in 1791, before going to the south to visit the Vesuvian area and some places near Rome in the Spring. Then, he returned to the Venetian Republic in June in order to carry out field trips in better weather conditions in the Euganean Hills and other places in the Veronese and Vicentine areas.

One target of these travels could be certainly considered as part of a long-standing tradition: in fact since the sixteenth century, the Vesuvian volcanic area had been extensively visited by a good number of British travellers (Capuano, 1994): among them, there were also some Irishmen such as the philosopher George Berkeley (1685–1753), who was in Naples in 1717 during an eruption of Vesuvius (Gasparini and Musella, 1991, pp. 207–208). Moreover, between the seventeenth and the eighteenth centuries, an interpretative model for travelling in Italy was elaborated by the British travellers (Comparato, 1979): the fashion of the Grand Tour among the English nobles drastically increased the number of trips made to Italy for different purposes. The travellers were mainly interested in art, but also in agricultural techniques or in aspects of manufacture and trade.

However, in the second half of the eighteenth century, the Grand Tour also included many scientific researches in the field for examining various aspects of nature (Stafford, 1984). A typical case of a Grand Tour with both artistic and geological objectives is given by the travels in Italy of Frederick Augustus Hervey (1730–1803), Bishop of Derry and Fourth Earl of Bristol, between the 1760s and the 1790s. He visited Vesuvius for the first time in 1766 and explored mountains and hills around Vicenza in the Venetian Republic and other places in northern Italy during the 1770s (Childe-Pemberton, 1925; Herries Davies, 1978a, p. 84).

Graydon's itinerary in Italy was perhaps inspired by Hervey's Grand Tour, because if a stop in Naples for an excursion to Vesuvius was considered almost obligatory for a scientist-traveller in Italy, the specific interest in the geological features of the Venetian area and especially in the remains of the local extinct volcanoes was surely less common. The writings published in the 1770s by two scholars, the British Ambassador in Naples Sir William Hamilton (1730–1803) and the British Minister Resident at Venice John Strange (1732–1799), suggested to Graydon which areas to explore (Graydon, 1794, p. 284). Hamilton in fact wrote several papers about Vesuvius and its eruptions from the 1760s to the 1790s and published a volume of 'Observations' on Vesuvius (Hamilton Sir W., 1772), while in the early 1770s, Strange investigated the groups of prismatic basaltine columns and other volcanic rocks in the Venetian State (Strange, 1775). The latter also had a significant correspondence with several Italian naturalists (Ciancio, 1995a). Hamilton and Strange gave impetus to the researches on the Continental volcanoes within the British scientific community (Porter, 1977, pp. 161–162). Both of them met Hervey during his travels in Italy and guided him on excursions, respectively, to Vesuvius and through the Venetian territory.

Graydon also had the opportunity to meet Hamilton in Naples and went up Vesuvius with the best local guide recommended by the British Ambassador himself (Geol/TCD/Mun/Graydon/1, p. 3). We do not know if Graydon also met in Rome the young Irish antiquarian Charles O'Conor (1764–1828), grandson of a distinguished member of the Royal Irish Academy. O'Conor was in Rome as a student of the Irish College during the 1780s and came back to Ireland after 1789 (O'Conor, 1949, p. 336). He was employed by the Royal Irish Academy in 1786–88 to buy books and research and copy Irish manuscripts in the Vatican libraries; however, after his return, he also presented to the Academy some specimens of Italian fossils together with a catalogue, neither of which have been found to date (Anon., 1794, p. 1; R.I.A. Council Minutes, vol. 1, pp. 79, 98, 109, 114, 125; O'Raifeartaigh, 1985, p. 19).

Graydon's collection of fossil fish from Bolca

Early in 1791, in January, Graydon spent a short time in Verona where he was shown a collection of fossil fish from Bolca. It made quite an impression on him, as he resolved that if he should return to the area, he would go to the quarry from where they came and would try to understand and explain the conditions of their formation. He returned in June that year and spent four days at Arzignano (west of Vicenza) from where he made collecting forays into the surrounding hills.

In Arzignano, the Irish traveller was given accommodation by the Paduan abbot Alberto Fortis (1741–1803), who perhaps guided him personally in some excursions in the Euganean Hills and in other districts of the Venetian territory (Graydon, 1794, p. 282). John Strange had left Venice in 1788, so Graydon addressed himself to Fortis, who had had a significant scientific relationship with Strange and had also followed Hervey in his travels of 1771 in Dalmatia and Italy, including Vesuvius and the Venetian territories (Ciancio, 1995b). The researches carried out by Fortis on the geology of different places in the Vicentine, Veronese, and Paduan areas were well known by Graydon, but he noted the difficulty even in Italy in procuring many writings of the Italian naturalists, such as Giovanni Arduino (1714–1795), which 'though written some years back, would form a most useful collection of accurate local descriptions and observations' (Graydon, 1794, p. 282).

Graydon undertook his geological excursions usually with the guidance of local peasants 'well acquainted with the country' and on one such trip he paid a visit to the quarry near the village of Bolca (at the top of the valley of Alpone, some 23 km north-east of Verona) where he collected a number of fish and plants for himself: he asked some of the people usually employed in working the quarry to bring their tools and show him their manner of operation. Graydon hoped to see for himself some fish opened in the stone. Unfortunately, the day chosen for the visit was very wet and only two or three workmen agreed to follow the Irish traveller, but from them, he got all the information he wanted. 'I spent nearly three hours on the spot'—Graydon writes—'during which time I not only saw some fish, as well as a few remains of marine vegetables found by the men, but had the pleasure to find some myself, opening with my own hands stones which contained them' (Graydon, 1794, p. 285).

Generally, the quarry workers sold the fossil fish, which often were extracted in many pieces, to dealers in Verona, who pieced, glued, and trimmed them prior to selling. Apart from collecting some specimens, Graydon also purchased several fish specimens from some Veronese dealers, whose names have not been traced. However, he was duped by one who sold him a fine specimen which later was recognised as containing the head of one species and the body and tail of another (Davis, 1888). The specimen, which is in the National Museum of Ireland (NMING: F12984), has the head of *Sparnodus macrophthalmus* Agassiz and body of *Seriola prisca* (Agassiz), and is an early example of a forged fossil.

This material was to become the subject of a paper, written after his return to Ireland and read by Graydon at a meeting of the Royal Irish Academy on the 1 March 1794, which was published the same year in the *Transactions of the Royal Irish Academy* (Graydon, 1794; Figure 3.1). In the first part of this paper Graydon gave a short introductory account of the lithology of Monte Bolca (the Bolca area), describing his fieldwork and the manner of working of the quarry workers in looking for the fossils, 'by detaching from the face of the quarry moderate sized blocks, which are then drawn out, set on edge, and quickly split with sharp-sided hammers or wedges'. He also noticed that besides the principal quarry, there were 'two or three others of the same stone' which had been recently discovered but never excavated because of the poverty of the local people (Graydon, 1794, pp. 286–287).

Concerning the fossil specimens collected in Bolca, Graydon acquired samples of some of the best-known Eocene fish: *Mene rhombea* and *Sparnodus elongatus*. While Graydon does not give precise details as to where he collected the specimens other than to state that they came from 'Bolca', they all are preserved in a fine-grained, buff-coloured limestone which is characteristic of that presently exposed at the quarry called La Pesciara. In the paper the fish are illustrated in a pair of lithographs (Figure 3.2).

Graydon's collection of fish and plants from the Eocene of Bolca numbered only eight specimens (part and counterpart are herein counted as the same specimen) but nevertheless was important from an Irish perspective as it was certainly

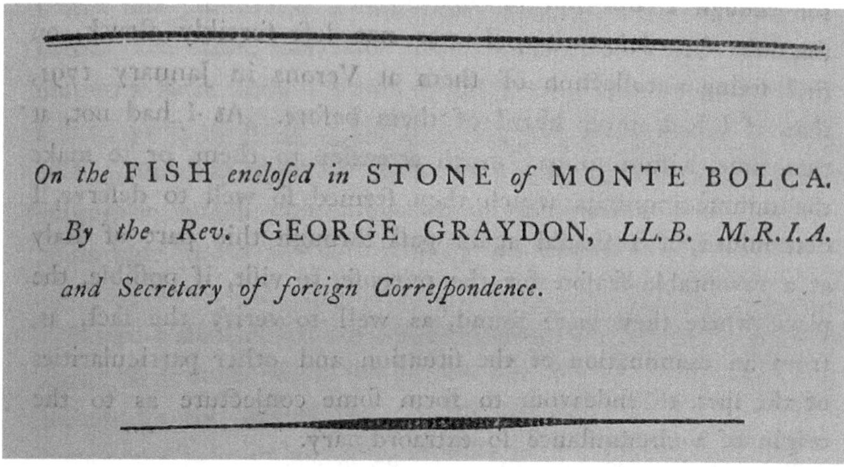
On the FISH *enclosed in* STONE *of* MONTE BOLCA.

By the Rev. GEORGE GRAYDON, *LL.B. M.R.I.A.*

and Secretary of foreign Correspondence.

Figure 3.1 Title of Graydon's 1794 paper (Geological Museum, Trinity College Dublin).

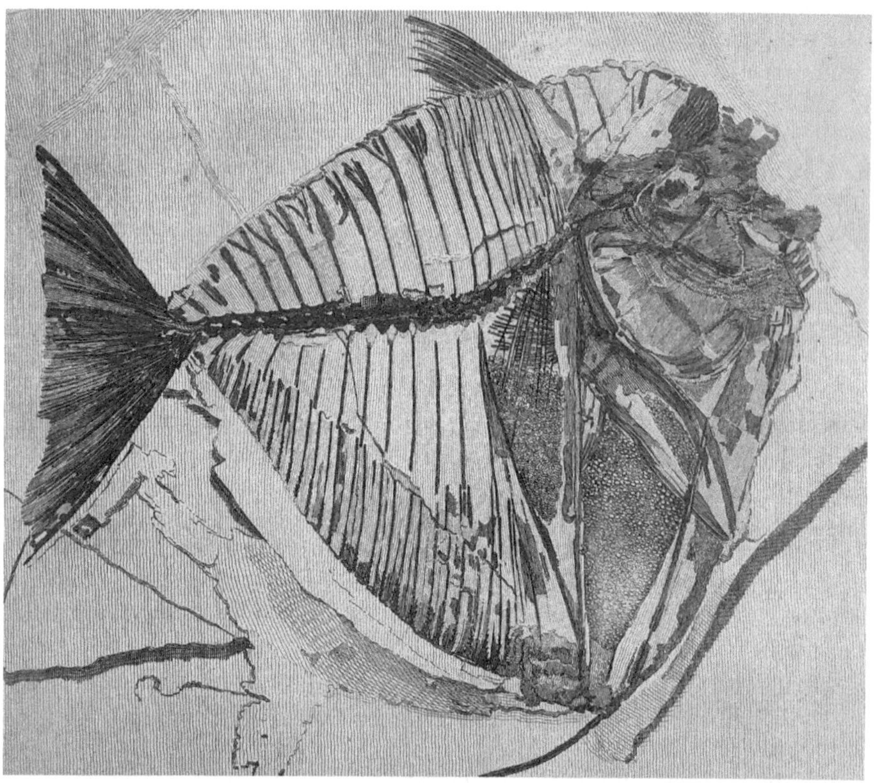

Figure 3.2 *Mene rhombeus*, an Eocene fish collected by George Graydon as illustrated in his paper published in 1794 (Geological Museum, Trinity College Dublin).

the first time that such material would have been seen or heard of in Ireland. Prior to being deposited in the museum of the Royal Irish Academy, the collection was carefully catalogued. The catalogue entry pertaining to the Bolca collection is given below, according to the manuscript catalogue kept in the Department of Geology, Trinity College Dublin (Geol/TCD/Mun/Graydon/2, pp. 104–105).

Class 4th
Substances of the Animal, Vegetable, and Mineral Kingdoms, connected remotely with and tending to Illustrate the Natural History of Volcanoes, and of supposed Volcanic Districts.

M1. Remains and Vestiges of Animals

No.

1. Fossil fish of Monte Bolca as described in the fifth volume of the transactions of the RI Academy and this specimen represented in the first plate –
2. Two small fish of Do, as the first plate—These are both single specimens.
3. Duplicates exactly corresponding containing on each side the impression and remains of the same fish as in plate 2nd.
4. Duplicates of part of a Marine Plant of Do, as in the 3rd plate—This was taken by myself at Bolca. The former was purchased by me at Verona –
5. Part of small fish of the head is wanting and there is a straight vein of thin blue that crosses the stone/just where the extremity of the Head is cut and of the colours of the laminae on each side of it do not exactly correspond as seen on the Edge it is probable that there was a [settlement] in the stone and that it was fractured and part of it sunk lower than the other carrying part of the fish in it and the Edges were afterwards cemented by spar so/as that the lie of each did not correspond. This is engraved in the 3rd plate and was found by me –
6. Duplicates of the impression or rather stain in brown of a leaf which resembles that of a Land tree or Plant that is probably Marine. It has left no dent of ribs or stem as it would probably have if they have been hard or prominent like those of land it is engraved in the 4th plate found by me –
7. A small fish, single, of the Spratt or Anchovy kind—probably as in the 4th plate and also taken by me at Bolca –
8. This is a single specimen and the largest that I brought Home. The brown Animal Matter is thick and the Bones well expressed in it. I bought it at Verona.

Modern identifications of Graydon's fishes with National Museum of Ireland catalogue number (NMING):

M.1.1 (Graydon catalogue entry number): *Mene rhombea* (Volta, 1796), NMING: F20639.

M.1.2: *Myripristis stenopteryx*, NMING: F20631.

M.1.3: *Sparnodus elongatus* Agassiz, 1835. Part and Counterpart, NMING: F16207/A&B.
M.1.4: possibly *Pterigophycos spectabilis* Massalongo, 1858. Part and Counterpart, Missing.
M.1.5: Small fish, Missing.
M.1.6: possibly *Protorchis monorchis* Massalongo, 1859. Part and Counterpart, Missing.
M.1.7: *Clupea catopygoptera* (Woodward, 1901). NMING: F16050.
M.1.8: head of *Sparnodus macrophthalmus* Agassiz, 1835 and body of *Seriola prisca* (Agassiz, 1834), NMING: F12984.

Studies of the Bolca fishes by Jacques Blot, Lorenzo Sorbini and others have been carried out since the 1960s, under the auspices of the Museo Civico di Storia Naturale, Verona. These authors have published important taxonomic revisions of the fauna mainly in the Museum's serial publications *Studi e Ricerche sui giacimenti terziari di Bolca* and *Memorie del Museo Civico di Storia Naturale di Verona* (listed in Solinas, 1978; Sorbini, 1981, pp. 130–132; Wyse Jackson, 1994a, p. 103). We have adopted these studies as a 'standard source' for the current names of the Bolca fish. The fish are found within a fine-grained laminar limestone, which is interbedded with a coarse-grained massive limestone containing a molluscan fauna. Five levels of the fish-bearing limestone have been noted within a total limestone thickness of 19 m at La Pesciara at Bolca (Blot, 1969, p. 27; Siliotti, 1979, pp. 26–39; Sorbini, 1981, pp. 53–55). The fish are presently dated as being between Lower and Middle Eocene in age (Siliotti, 1979, p. 39, fig. 24). According to the stratigraphical studies of the area, this limestone series represents the fossiliferous horizon of La Pesciara, which has been placed at the base of the broader horizon of Monte Postale. This latter horizon is extremely fossiliferous and includes limestones, volcanic breccias and tuffs, lignites, and clays (Siliotti, 1979, pp. 26–27). The precise level from which Graydon's specimens were collected has not been identified.

Graydon and the nature of the fossil fish from Bolca

The fossiliferous site of Bolca has been known in Italy since the sixteenth century and was studied mainly by distinguished Italian scientists, particularly during the eighteenth century (Solinas, 1978, pp. 7–9; Sorbini, 1981, pp. 14–41). During his stay in the Venetian State, Graydon tried to collect not only specimens but also all the available information about Bolca, particularly regarding the formation and the unusual concentration of its fossil fish.

He soon understood that among many Italian and non-Italian naturalists, such as Arduino, Fortis, and Strange, the predominant view on the lithology of the area was that it was covered with 'productions' of extinct volcanoes. As for Graydon, his own observations in the field demonstrated that the 'supposed compact lava of the Vicentine and Veronese [...] in short almost perfectly resembles our basaltic stones of the country of Antrim' (Graydon, 1794, p. 284).

The land containing the quarries of Bolca had been the property of Vincenzo Bozza, a pharmacist-chemist from Verona who had in 1790 a renowned cabinet containing about 700 specimens of fossil fish collected in the previous twenty years (Pomian, 1986, p. 46). While Graydon was in Verona in 1791, the elderly Bozza sold his properties in Bolca and his cabinet of fossils to Giovanni Battista Gazola (1757–1834), a gentleman who already had a fine collection in Verona (containing up to 1,200 specimens in 1792) which included several fossil fish which were not included in Bozza's cabinet. The Irish scholar received directly from Gazola, who was a good friend of Fortis too, a list of these fossils from the Gazola collection and the printed catalogue of the rarest specimens of fossil fish from Bolca in Bozza's cabinet. These were put together and printed in the original Italian version at the end of Graydon's paper of 1794. During the same years, Bozza's cabinet and catalogue of fossil fish were known also to the Scottish scientist John Walker, Professor of Mineralogy in Edinburgh, who listed some specimens contained in the cabinet of the Veronese collector in the manuscript memoir 'Upon the Ichthyolites of Monte Bolca' (Walker, 1793).

Bozza, who followed some suggestions given by Fortis, and later Gazola, recognised that the fossil fish from Bolca were very similar to those still living in the tropical seas (Fortis, 1786; Bozza, 1788; Gazola and Tommaselli, 1794). In fact, their classification, also published by Graydon, subdivided the fossil fish into six classes ('ordini'), based on their distribution in different seas (European, Asiatic, African, South American, North American) and in fresh water (Graydon, 1794, pp. 312–317). The theory of these Veronese naturalists, who also included Giovanni Serafino Volta (1754–1842) from Mantova, was that Noah's Flood caused the transportation of the fishes from the tropical seas to Bolca, once also covered by seawater (Sorbini, 1981, pp. 31–32). Graydon was well informed about the ideas published by Vincenzo Bozza, as well as the similar position of Gazola (Graydon, 1794, pp. 288, 292), but in his paper, he did not mention the contributions to the debate on the fossil fish of Bolca made by Giovanni Serafino Volta (Graydon quoted only Volta's handbook *Elementi di Mineralogia*, published in 1787), the Roman abbot Domenico Testa (1746–1825), and Alberto Fortis (Volta, 1789, 1794; Testa 1793a–c; Fortis, 1793). His lack of knowledge about the continuation of the debate, which became a real controversy in 1793 between the theory of the Veronese naturalists and the different views of the abbots Testa and Fortis (Pomian, 1986, pp. 49–51), can perhaps be explained by Graydon's absence from Italy since the end of 1791. In Ireland, it was of course very difficult to obtain these very specialised scientific Italian publications probably printed in a limited quantity, and Graydon tried to find more updated information on Bolca in the literature available in Dublin or in the books he found in Italy, such as the *Gabinetto Mineralogico del Collegio Nazzareno*, published in Rome by Giovanni Vincenzo Petrini in 1791–1792 (Graydon, 1794, p. 292).

Testa in particular had stated that the majority of the fishes observed in Bolca were not tropical and the cause of their sudden death and concentration was not due to the Deluge but to a violent submarine volcanic event. Fortis did not accept the Deluge as a cause of this phenomenon, but he also refused to accept Testa's volcanic theory and suggested instead the slow deposition of sediments in the seawater as the cause of the fossilisation of the fish (Sorbini, 1981, p. 33).

Graydon probably did not know the works of Testa when he wrote in the second part of his paper of 1794 a detailed explanation of the causes which determined the formation and the concentration of the fossil fishes in Bolca. Certainly, when he personally illustrated for the first time the outlines of his hypothesis to Gazola in Verona, immediately after the visit to Bolca in June 1791, the Veronese collector 'seemed wholly unacquainted with any thing similar' (Graydon, 1794, p. 292).

The examination of the collected specimens had soon convinced Graydon that the forms of the fish were preserved 'uncommonly perfect' and the dry remains of the animal bodies were in such a state 'as almost authorize their being called fish-mummies'. Consequently, this evidently perfect preservation was considered by the Irish cleric as 'certain proof' that the fishes were enclosed in the matter which surrounded them quite shortly after their death, without having suffered 'general putrefaction'. According to Graydon, the enclosing matter was a 'whitish, yellowish or bluish grey' calcareous stone 'of a schistose structure, capable of being split into flags or laminae of various thickness and dimensions', which was also called 'marle or marley schist' (Graydon, 1794, p. 289). At the beginning, it was suddenly spread through the water 'in a very fine and pulverulent state' and probably was the cause of the death of the fishes (Graydon, 1794, pp. 294–297). All these operations took place 'in or under water' of an ancient sea, but Graydon seemed to be rather far from adopting the theories of the Veronese naturalists about the fishes being transported from the tropical seas to Bolca and the possible influence of Noah's Flood (the latter is never mentioned in his paper).

The first general conclusion was in fact that 'the sudden diffusion of lime in the water in which these fish happened to be, and its consequent deposition, was the immediate cause of their enclosure, and the origin of this curious quarry' (Graydon, 1794, p. 301). The fishes were supposed to live in a sea much nearer to Bolca than in the present (if not wholly covering the highest hill of the area as well) and only a great change in the environmental conditions could have determined their death and enclosure in the limestone. The observation of rocks and strata around Bolca suggested to Graydon a possible answer about the cause of the hypothetical sudden diffusion of lime in the water: because the limestone had been burned and calcinated in order to be spread 'in a pulverulent state' through the water, the cause could only be fire. And the immense quantity of this 'unexpected projection' of lime, clearly indicated by the thickness of the limestone strata, could be due to a volcanic eruption, as Graydon himself had already verified on Vesuvius. In fact, this volcano had often erupted marble and other calcareous stones, 'in various degrees of calcination', and some of them were contained in the collection of Italian rock specimens donated by Graydon to the Royal Irish Academy.

The problem of the great number and variety of fossil fish found in Bolca was left unsolved by the Irish scholar, who rejected as a possible cause 'submarine volcanic commotions' which could have concentrated so many fishes in one place. According to Graydon, some 'un-common cause' certainly lay behind the assemblage in Bolca of fishes from different seas and rivers. It is interesting to note that this supposed presence of freshwater fishes represented for him the proof of the existence of mountains and hills at a much higher level than the sea, of course, including volcanoes (Graydon, 1794, pp. 302–303).

Graydon also gave a short and general account of the lithology of the hill of La Pesciara:

> The whole of the hill, as far as I could observe, seems to be composed of similar, or at least of argillaceous matter, except the quarries in which the fish are found, which are calcareous, and lie at about half a mile from the summit. Besides the dissimilarity of these to the other materials of the hill, it is further important to remark that they do not form a continued stratum, but lie in wholly detached and distinct masses.
>
> (Graydon, 1794, p. 284)

The Italian naturalists, who debated at the end of the eighteenth century about this fossiliferous site, even if their attention was mainly focused on the fascinating question of its origin, also noted the rocks associated with the fossils. The interest of the scientists in the lithology of this fossiliferous site had notably increased in the second half of the eighteenth century, since the time of the first known geological sketch of the Bolca area, drawn by Giovanni Arduino in 1740 and published by Giovanni Giacomo Spada in 1744 (Sorbini, 1981, pp. 27–30; Vaccari, 1993, pp. 34–36).

The last part of the 1794 essay was dedicated by Graydon to the explanation of the geological features of the Bolca quarries. The conformation and composition of the calcareous masses, lying at different heights and inclinations, suggested to Graydon that their original position must have been different, that is to say, horizontal as part of a large continuous calcareous stratum. This was probably the bottom of the ancient sea and its extent must certainly have been very wide: the quarries of Bolca showed only some fragments or portions 'completely disjoined from all the connections with their native bed'.

As for the cause of this situation, Graydon assumed that after its consolidation, the original stratum was violently broken up by 'subterraneous fire and explosion' with an eruptive force never experienced in the history of active volcanoes. This event destroyed every trace of the original stratum all around the quarries, as most of the territory of Bolca and its surroundings were covered, according to Graydon, with 'argillaceous materials' (considered similar to the basalt of Antrim), supposed lavas and ejecta of ancient submarine volcanoes.

The question of the origin of basalt, which clearly emerged from these last considerations, was deliberately confined to a few lines, where Graydon diplomatically avoided taking a public position in the controversy, although the volcanic theory of the formation of basalt in the Venetian area, already suggested by scientists such as Fortis, Arduino, and Strange, seemed to be confirmed and reinforced by the content of this essay on the Bolca fossil fish (Graydon, 1794, pp. 304–308).

In conclusion, Graydon summarised the results of his geological studies in Bolca by stating that

> the remote existence of two distinct areas of great volcanic explosion in this place, one of which at least seems decidedly to have been submarine; the first, when the fish were caught and inclosed in calcined calcareous matter; and the second, after the complete concretion of this matter into stone, when its horizontal stratum was torn up and dispersed
>
> (Graydon, 1794, p. 309)

It is interesting to note that, within this volcanistic view of the origin of the fossiliferous site, Graydon showed in his paper a knowledge of a memoir by Ermenegildo Pini (1739–1825), which strongly supported a sudden, fast, and extraordinary flood as an explanation of all the phenomena which occurred in Bolca, including the unusual concentration of the many varieties of tropical fish (Pini, 1792, pp. 460–464). According to Pini, the activity of a submarine volcano could also be accepted (because of the existence of many volcanic rocks in the area), but only as a secondary agent which acted contemporaneously with the flood.

On the other hand, Graydon did not mention other former volcanistic interpretations of the origin of the Bolca quarries, which in some way partially anticipated his own conclusions on the two distinct volcanic activities. In 1740, Anton Lazzaro Moro (1687–1764) had written that all the fishes at Bolca were buried directly by the lava that had erupted from a newly formed volcano (Moro, 1740, pp. 368–370), while about thirty years later, Giovanni Arduino stated that the calcareous strata observed in the quarries were probably part of the original bottom of the ancient sea, detached and pushed to different heights and inclinations by volcanic forces (Arduino, 1782 [1769], p. 163; Vaccari, 1993, p. 300). A similar hypothesis was also suggested in the writings of the Veronese scholar Scipione Maffei (1675–1755) in 1735 and 1747 (Sorbini, 1981, pp. 25–29; 1985).

In spite of this, Graydon's attempt to explain some of the phenomena linked to the fossiliferous site of Bolca remains of interest for the history of geology and palaeontology in late eighteenth-century Italy and Ireland. The interest aroused within the Royal Irish Academy upon the reading of Graydon's paper on Bolca was also shown by a letter about the preservation of the solid parts of the fishes in the limestone, addressed to Graydon by the Rev. Hugh Hamilton and read in the Academy on the 5 April 1794 (Hamilton, 1794).

Graydon's collections in Ireland

On his return from Italy late in 1791, the Minutes of the Royal Irish Academy Council tell us that Graydon donated a 'curious collection of specimens' to the

Academy museum, for which the Council, on the urging of Richard Kirwan, granted him £20 in February 1792. This collection probably contained some material collected earlier in 1791, but the bulk of his Italian collection of lavas and other rocks and fossils, including the Bolca fish and plants, was not donated until May 1794, after his paper on the Bolca, fish had been read and accepted for publication (R.I.A. Council Minutes, vol. 1, pp. 223, 228, 291). He had spent three years sorting the collections into a suite of the finest material for the Academy and a set of duplicates for himself or for donation to others. Graydon received generous thanks from the Academy and earned the highest praise from Kirwan for the assiduous nature with which he assembled the fine collection (Kirwan, 1794, p. xv). When published, the paper amounted to 37 pages and contained 4 plates. In the copy, we have seen the order of the first two plates is reversed relative to their positions cited in the catalogue entry.

At about this time, Graydon gave a Mr Rainy (or whom no more is known) two hundred and ten specimens and may also have passed some specimens to White Watson, who by coincidence was an acquaintance of Mr Palmer who had spent a week with Graydon on Vesuvius (Wyse Jackson and Vaccari, 1997).

In March 1801, the Dublin Society requested, with Graydon's consent, that his collection of fossil fishes and volcanic specimens be transferred to their Museum, which would allow it to be seen by the general public (R.I.A. Council Minutes, vol. 2, p. 125; Anon., 1862, p. 277). The Dublin Society had the means to pay a Professor of Mineralogy, and had only a few years earlier in 1792 had purchased the important Leskean Collection of over 7,000 geo-mineralogical specimens (Vaccari and Monaghan, 1993). Both these factors increased the importance of their Museum and had reduced the necessity for the Royal Irish Academy to maintain a geological and mineralogical collection. Nevertheless, it continued to acquire Irish antiquities and sixty years later received a donation of such material from the Royal Dublin Society (Anon., 1862, p. 277).

The Council of the Royal Irish Academy agreed to the Dublin Society's request shortly afterwards but stipulated that the Graydon collection should be kept separate from the Leskean collection and, second, if it was removed from public display, it should be returned to them (R.I.A. Council Minutes, vol. 2, p. 128). The Dublin Society collections were further enhanced by Graydon's donation of a collection of fossils from the Mediterranean. For his efforts in arranging the transfer, he was elected an Honorary Member (Anon., 1801, p. 122). In 1804, after his death, his widow donated a small suite of specimens from the north of Ireland to the Dublin Society, for which she received its thanks and a gold medal (Anon., 1804, p. 34), and sold Trinity College, Dublin the residue of his Italian collection of 1,370 specimens for the large sum of £100 (TCD/Mun/P/4/87/45).

Responsibility for the Museum of the present-day Royal Dublin Society was taken over by the Government in 1877, after which it became known as the Museum of Science and Art, which was later to be retitled the National Museum of Ireland (Monaghan, 1992).

Today, only some of the fish specimens (but none of the plants) from Bolca can be found in the collections of the National Museum of Ireland. The volcanic specimens and the other fossils from Roncà and other localities have not surfaced. As much of the geological collections were packed away in 1961, it is possible they are still in crates. Since 1982 approximately 70% of the crated material has been unpacked, fully documented, and repacked (Monaghan, 1992). In Trinity College, some of Graydon's specimens were incorporated in the mineralogical collections in 1850, while earlier a small number had been given to the Swiss geologist Jean-Francois Berger in 1811. The latter were transferred to the Geological Society Museum and later to the Natural History Museum in London. One specimen of a leucite-bearing basalt is still extant in London (Wyse Jackson and Vaccari, 1997). In Trinity College, however, there is no trace of the majority of the volcanic specimens: these were probably disposed of in 1954 when the Geological Museum was reduced in size (Wyse Jackson, 1992b).

Two ledgers containing copies of Graydon's catalogues and geological writings survive in the Department of Geology in Trinity College (Geol/TCD/Mun/Graydon/1; Geol/TCD/Mun/Graydon/2) and are important as they give an indication of the size and content of the collection and allow us to assess its importance. Unfortunately, the originals that were formerly in the Dublin Society and may have been transferred to the National Library of Ireland cannot now be found.

Manuscript sources

Department of Geology, Trinity College Dublin:

Geol/TCD/Mun/Graydon/1—Ledger containing a copy of the diary of Graydon's travels in the Vesuvian area and his various observations on volcanoes and their products. Watermark date 1803.

Geol/TCD/Mun/Graydon/2—Ledger containing a copy of the catalogue of collection of rocks, minerals and fossils made by Graydon in northern Italy and in the Vesuvian area. Watermark date 1802.

Linnean Society, London:

Walker J., 1793—Memoir upon the marble quarries of Carrara & upon the Ichthyolites of Monte Bolca in the Veronese. (Soc. Prom. Nat. Hist. Manuscript, 7 leaves)

Manuscripts Department, Trinity College Dublin:

TCD/Mun/P/4/87/45—Note outlining that a £100 draft was received for paying Mrs Elizabeth Graydon for the collection of minerals made by her late husband and sold to Trinity College. Dated 23rd August 1804.

National Library of Ireland:

Ms 16964—Account of the administratorship of Elizabeth Graydon. 1803–1807.

Royal Irish Academy:

R.I.A. Council Minutes—Minutes of the Council of the Royal Irish Academy, vol. 1, 1785–April 1797; vol. 2, 1797–1809.

R.I.A. General Meetings Minutes—Minutes of the General Meetings of the Royal Irish Academy, vol. 1, April 1785–November 1826.

4

RICHARD KIRWAN (1733–1812): CHEMIST AND GEOLOGIST

Irish Chemical News 12(1) (1998), 36–39.

Introduction

Richard Kirwan (Figure 4.1), perhaps the finest scientist produced by eighteenth-century Ireland, is now remembered largely for his eccentricities, for his opposition to the ideas of James Hutton, and for his championing of the discredited phlogiston theory. A polymath he published many books and lengthy papers on chemistry, meteorology, and geology. Although many of his ideas were not accepted by his contemporaries, nevertheless they were important in that they prompted the publication of other theories and ideas that advanced New Chemistry and the modern understanding of geological processes.

Origins of a scientific life

Richard Kirwan was born on 1 August 1733 at Cloughballymore in County Galway, Ireland into a prominent Catholic landowning family, one of the fourteen tribes of Galway (Donovan, 1850; McLaughlin, 1939–1940; Dixon, 1971; Scott, 1973, 1979; Burns, 1985). The second son of four born to Martin and Mary (née French), he showed a remarkable talent for learning from an early age. After schooling, he was sent, at the age of seventeen, to Poitiers in France where, much to the consternation of his mother, he began to purchase books on chemistry, both in English and in French, which laid the foundations for much of his later work. He later moved to St Omer in order to study for the priesthood. There he excelled at classics and was appointed Professor of Humanities.

In 1755, when Richard was twenty-two years old, his elder brother Patrick was killed in a duel with a Mr Brereton, a porter or Usher of the House of Commons, and he succeeded to the family estates at Cregg, County Galway and an income of £4,000 a year. Soon afterwards in 1757, he married Anne Blake, daughter of Sir Thomas Blake of Menlough Castle. It has been reported that Kirwan was jailed on his wedding night on account of his bride's debts. After his release, he moved to Menlough Castle, which he and his wife shared with his mother-in-law. There he spent much time studying and working in his library and purpose-built chemical laboratory which his wife's mother disapproved of (Burns, 2003). His wife bore

DOI: 10.4324/9781003475552-7

Figure 4.1 Richard Kirwan 1733–1812 (From Donovan, 1850) (Geological Museum, Trinity College Dublin).

him two daughters, Maria Therese and Eliza, but died at an early age in 1765, after which the girls were fostered by Kirwan's mother-in-law.

Maria Therese married Lord Trimleston in 1793 and Eliza married Colonel Hugh Hill in 1792.

In 1761, he moved to London to study law, and was later called to the Irish Bar in 1766. In order to pursue this profession, it was necessary to be of the Protestant faith, to which Kirwan had conformed some years earlier. Perhaps, characteristically for one who was ill-at-ease as to what his life's work should be, he abandoned law in 1768 in preference for a life engaged in scientific study and endeavour. He left London in 1772 and returned to Cregg Castle, Co. Galway via Dublin where he spent the next nine years learning Greek, other European languages, and assembling a fine library. He was also an accomplished musician and assisted Edward Bunting (1773–1843) in his compilation of Irish airs.

By the age of forty, Kirwan had equipped himself, both mentally and physically with the apparatus for his intended future studies in chemistry and mineralogy and geology. In a paper published since the first appearance of this chapter, Sally Newcomb (2012) has provided an in-depth assessment of Kirwan's scientific output.

London sojourn: Kirwan's work in chemistry

In 1777, he returned to London where he soon became involved in several societies both influential such as the Royal Society of which he became a Fellow in 1780 and the less so, such as the Chapter Coffee House Society. He was unfortunate in being a victim of privateers on the high seas when the ship the *Duke of Gloucester,* which was transporting his library to London, was captured by the American ship the *Pilgrim,* whose captain was ironically Thomas Kirwan, who was descended from those of Cregg Castle. This library is now largely shelved in the Salem Athenaeum in Salem, Massachusetts in the United States (Burstyn, 1960). Kirwan accepted the loss and rebuilt his library, not an easy task.

In London, Kirwan resided at 11 Newman Street, off Oxford Street which became an important centre for Wednesday evening conversaziones for scientists such as Joseph Priestley, Sir Joseph Banks, and Henry Cavendish. It was during the next decade that Kirwan carried out much of his work on chemistry. For his work on chemical affinity, he was awarded the Copley Medal of the Royal Society in 1782. Other important contributions to chemistry included his titrimetric determination of iron with ferrocyanide, and his tabulation of specific heats based on water (Burns, 1977).

In 1787, the year that he returned to Dublin and took up residence at 6 Cavendish Row, his most important work in chemistry *An Essay on Phlogiston and the Constitution of Acids* was published; a second edition appeared two years later. In this work, he advanced the theory that phlogiston was a constituent of all combustible substances, which when burnt lost phlogiston. This theory was challenged by the French chemist Lavoisier and Kirwan's fellow countryman William Higgins, who showed that there was an increase in weight on combustion of metals and that oxygen was absorbed in the process. Kirwan's *Essay* was translated into French by Lavoisier's wife with notes refuting his ideas added by the French chemist. Eventually, Kirwan accepted Lavoisier's findings. It was, as William Davis has suggested, Kirwan's anti-phlogiston stance that led others to formulate and advance the theories that became known as the New Chemistry.

Dublin scientific circles

From the mid-1700s, chemistry and mineralogy became a focus for serious scientific study. Ireland, and in particular Dublin, became an important centre for such studies, ranking close in importance to Edinburgh. The reasons for this were the establishment of various institutions such as the Royal Irish Academy, the Dublin Society, and the developing international reputation of Irish scientists. Kirwan was the foremost scientist in Dublin from the late 1790s until his death, a period that coincided with his presidency of the Royal Irish Academy from 1799 until his death thirteen years later. This body was founded in 1785 (Kirwan was a founder member) for the study of sciences, polite literature, and antiquities, and it elected its members from the noble, clerical, and academic circles of the day and rapidly built up an important library and museum. Many important papers on the subject were published in the Academy's *Transactions*. This journal was exchanged with similar learned institutions in Europe, which allowed for a rapid diffusion and exchange of ideas and no doubt debate.

The Dublin Society was established in 1731 to promote agriculture and sciences in Ireland, by organising county statistical surveys which were promoted by Kirwan, and through its museum which thanks to him, acquired in 1792 the important Leskean collection of 7,331 mineral specimens. The chemist William Higgins was soon afterwards briefly employed to organise this massive collection, which is still housed in the National Museum of Ireland. Kirwan was awarded a medal of Irish gold by the Dublin Society for his efforts. The presence of the collection in Dublin stimulated considerable geological activity by members who made mineral collections for themselves, many of which ultimately found their way into the Museum or other institutions. Kirwan recognised that the study of mineralogy was important for economic considerations: he was an Inspector of Mines, wrote two memoirs on mines in Ireland, and proposed that a Mining Board be established. In 1812, a number of his friends established the Kirwanian Society, which promoted many of his ideas, but which fell into abeyance in 1818 (McLaughlin, 1954).

Geological studies

Kirwan, the geologist could be accused of being an armchair researcher who was dependent on the printed word and his mineral cabinet which was said to be one of the finest in private hands. What has become of it is unfortunately presently unknown. This is possibly a fair assessment of his geological researches, but as a geological thinker Kirwan, in the later 1700s had almost no equal.

He produced two important books *Elements of Mineralogy* (1784, 2nd ed. 1794) and *Geological Essays* (1799) which went to a number of editions and were translated into several languages. Through these works, Kirwan's ideas were transmitted to many not only in Ireland and Britain but also on the Continent and in northern America.

The former was the first English-language text on the subject while the latter was published as a retort to James Hutton's paper of 1788, with which Kirwan found much fault. Hutton was a Scottish gentleman farmer who advocated that the nature of the geological record was cyclical and that rock genesis and erosion were an ongoing process that had taken place repeatedly many times in the past.

Kirwan abhorred the anti-biblical tenets of Hutton's ideas. He was particularly concerned with the lack of evidence presented by Hutton and said that the theory was at variance with geological knowledge at that time. Kirwan himself believed that earth history and structure could be explained with reference to the Bible and subscribed to the theory that all rocks were precipitated from some primordial fluid, and that these were later eroded and shaped by the waters of the Flood. He suggested that the general and specific attractiveness of particles caused differentiation in the early Earth. He believed that coal was the product of breakdown of granite and basalt which yielded bitumen which then aggregated into discrete layers. Galway Bay was, according to Kirwan, the site of an ancient granite body which was removed by the Flood (Gillispie, 1951; [Herries] Davies, 1969b; Dean, 1992). In the debate on the nature of basalt, the 'Neptunists', included Kirwan in their number (Wyse Jackson, 2000. Chapter 2 herein).

Flies and door knockers: final years

Towards the end of his life, Kirwan became celebrated for his eccentricities (Somerville-Large, 1975). He found it difficult to swallow and as a result always ate alone, dining on ham and milk. He received visitors on Thursdays and ordered Pope, his manservant, to remove the door knocker from the hall door of his house on Cavendish Row in Dublin every evening at 7 o'clock so as to stop further visits. He developed a paranoia about flies. He began to read religious tracts. He kept a pet eagle. He always wore a large foppish hat indoors and was terrified of getting cold. Lady Morgan once described how she was received by him on a warm evening and was surprised to find Kirwan wearing his hat as well as a cloak and shawl, surrounded by screens with a fire roaring in front of him. He died on 1 June 1812 in his eightieth year while attempting to starve a cold. The measure of his standing in Dublin can be gauged by the attendance at his funeral at which 900 persons followed his coffin as it was transported through the city. The route from Cavendish Row went through Sackville Street, Westmoreland Street, past Trinity College, up Dame Street, and crossed over to Capel Street before finally winding its way along Bolton Street and Temple Street. Kirwan lies in the former graveyard of St. George's Church, Hill Street, Dublin, which is now tarmacadamed and utilised as a football pitch for youths.

Unfortunately, only a handful of Kirwan's letters are extant, in Uppsala, Philadelphia, Birmingham, and London, and his correspondence with Guyton de Morveau has been published (Grison *et al.*, 1994). It is thought that most of his papers were burnt by members of his family after his death, and subsequently, barely any other materials have come to light. The philosophical and geological portion of his library was bequeathed to the Royal Irish Academy, while the remaining 3,000 volumes of his books and journals, including his personal run of the *Transactions of the Royal Irish Academy,* were auctioned after his death (Traxler-Brown, 1991).

Conclusions

Richard Kirwan was a prolific author who published eight books. Some appeared as multiple editions in English while four were translated into French, German, Spanish, or Russian. He also published over fifty papers in the fields of chemistry, geology, and meteorology, many of which ran to many tens of printed pages.

Kirwan has long been recognised as an important figure in the history of the chemistry and earth sciences, primarily for his work on phlogiston and as the catalyst for the publication of James Hutton's expanded theory and for his *Elements of Mineralogy,* the first English-language book of its type.

He received many honours: he was an honorary member of the academies of Berlin, Uppsala, Stockholm, and Philadelphia and was conferred with a Doctorate of Laws by the University of Dublin. He refused a baronetcy offered by Lord Castlereagh, the Chief Secretary for Ireland at the time of the Union.

Perhaps on account of his adoption of now discredited ideas, or his eccentricities in later life, or both, Richard Kirwan's work is not widely known or understood today. Only in the fields of chemistry and geology have some significant attempts been made by Scott (1973, 1979), Burns (2003), and Newcomb (2012) to assess the merits and sphere of influence of his work. Kirwan certainly deserves to be placed among the important shapers of the chemical and geological sciences during the eighteenth century.

5

WILLIAM HENRY FITTON (1780–1861) AND THE WOLLASTON MEDAL OF 1852

Geoscientist 8(10) (1998), 9.

During 1798 rebellion in Ireland, a young student of geology got inadvertently embroiled. William Henry Fitton (Figure 5.1) was arrested during those turbulent times while engaged in geological fieldwork collecting fossils in the Dublin region. His offence? He was carrying a geological hammer which was considered by the authorities to be a dangerous weapon. Although a classics scholar Fitton gained an interest in geology whilst an undergraduate. He graduated B.A. in 1799 from Trinity College Dublin which is the only constituent college of the University of Dublin, and following this, he spent a further number of years there, during which time he undertook geological study trips to Wales and Cornwall. He also determined barometrically the heights of the principal mountains in Ireland (Horner, 1862). In 1808, he moved to Edinburgh where he studied medicine, but also attended the lectures of Robert Jameson. He eventually gained the degree of M.D. from two universities: Edinburgh (1810) and Cambridge (1816). Some biographical accounts of Fitton credit him with the same degree from Dublin but a listing of graduates (Todd, 1869) shows that it was another William Fitton who gained this degree.

He settled in London for a number of years where he joined the Geological Society. With the Rev. Walter Stephens he co-authored an important memoir on the geology of Dublin, which was published following Stephens' death, in the Society's *Transactions* in 1811. An enlarged version appeared the following year. In 1812, he moved to Northampton where he practised medicine for eight years until he married Miss James, a lady of considerable means. He was thus able to abandon his profession and devote himself to scientific and more leisurely pursuits. He became more involved with the work of the Geological Society, serving on Council for two periods 1822–1830 and 1831–1846, Secretary 1822–1824, and President 1827–1829. During his term as President, he instituted the Society's *Proceedings* and held Sunday evening social gatherings for Fellows in his house (Wills, 1935). Roderick Impey Murchison who was Secretary for much of this

DOI:10.4324/9781003475552-8

Figure 5.1 William Henry Fitton, President of the Geological Society of London 1827–1829 (From Woodward, 1908; Geological Museum, Trinity College Dublin).

period found Fitton to be friendly but at times peppery and short-tempered, as well as being a lively debater, but who was unfortunately rigidly formal in the dealings of the Society. This resulted in a deterioration in relations between the two, which were only patched up thanks to the intervention of other Society members (Geikie, 1875).

Fitton carried out important work on Jurassic and Cretaceous rocks of southern England, distinguishing the Upper and Lower Greensand and the Gault Clay (Fitton, 1824, 1836). Of significant importance too were his lengthy reviews of contemporary geological writings—including those of A.G. Werner—which appeared in *Nicholson's Journal of Natural Philosophy* in 1813, and those of William Smith, William Buckland, Roderick Murchison, and Charles Lyell, published in *The Edinburgh Review* between 1818 and 1841. It was Fitton's review of Smith's map and books that allowed their significance to become generally known to the geological community at that time (Eyles, 1972), and it thus elevated Smith

Figure 5.2 Wollaston Medal, struck in Palladium, awarded to Fitton in 1852. (A) Obverse with profile of W.H. Wollaston; (B) Reverse with inscription enclosed in palm and laurel branches (Geological Museum, Trinity College Dublin).

into the rightful position amongst his peers. Is it coincidental that Smith should have received his only degree—an honorary LL.D. conferred at the British Association for the Advancement of Science meeting in Dublin in 1835—from Fitton's alma mater? ([Herries] Davies, 1969a).

On 20 February 1852, Fitton was awarded the Wollaston Medal (Figure 5.2A, B) by the Geological Society of London for his geological writings and research especially for explaining 'the differences between the Chalk and the Oolites, and their relations to each other' (Hopkins, 1852). In reply, Fitton touched on his friendship with Wollaston and stated the award 'will be prized, both by my family and myself'. (Fitton, 1852). According to Horace Woodward's history of the Geological Society, the medal was one of a batch struck in palladium and awarded between 1846 and 1860. Palladium had been discovered by William Hyde Wollaston in 1809, and the metal for the medals was extracted from Brazilian gold by the assayer Percival Norton Johnson (Woodward, 1908).

To date, eleven graduates of the University of Dublin have been awarded the Wollaston Medal. Of these four were once undergraduates: Fitton (B.A. 1799, medallist 1852), Robert Mallet (B.A. 1830, medallist 1877), Henry Medlicott (B.A. 1850, medallist 1888), and Christopher Hawkesworth (B.A. 1970, medallist 2012). One, William Johnston Sollas, conferred with an M.A. in 1884, received the medal in 1907. The remaining graduates were recipients of honorary degrees: William Smith (LL.D. 1835, medallist 1831), John Phillips (LL.D., 1857, medallist 1845), Sir Richard Griffith (LL.D. 1849, palladium medallist 1854), Herbert Harold Read (D.Sc. 1956, medallist 1952), Sir Harold Jeffreys (D.Sc. 1956, medallist 1964), and Philip Henry Kuenen (Sc.D. 1956, medallist 1970).

In 1997, Fitton's medal turned up in a bank vault in Canada where it had lain for nearly three decades. Owned by a collector of military medals who was selling his collection, it was seen by a similarly interested collector, who when he read accompanying documentation, recognised a link between a close friend of his and Fitton—both were graduates of Trinity College, Dublin. Soon afterwards, the medal was acquired by John Payne, a Trinity geology graduate of 1972, and generously presented by him to its Department of Geology. It is fitting that 200 years after William Fitton's arrest in County Dublin, the Geological Society's Wollaston Medal awarded to him, the only undergraduate of Trinity College to serve as president of that body, should find a permanent home with other items of Fittoniana in the Geological Museum of his alma mater.

6

WILLIAM SMITH AND IRELAND: SOURCES OF IRISH GEOLOGICAL INFORMATION ON HIS GEOLOGICAL MAP OF 1820

Earth Sciences History 35(1) (2016), 115–123.

Introduction

William Smith's 1815 geological map of England and Wales is a masterpiece of geological cartography and has been the subject of major study over a considerable period of time (Sheppard, 1915; Eyles and Eyles, 1938; Cox, 1942; Wigley *et al.*, 2007; Henry, 2014) and also Fuller (1995) who highlighted the significance of Smith's cross-sections. In 1844, Smith's memoirs were assembled and published by his nephew John Phillips, and since then, Smith himself and his labours have been the subject of continual assessment (Eyles, 1969a). These specifically include the evaluation of his stratigraphical ideas (Cox, 1948), his fossil collection and its dispersal (Eyles, 1967) and their reassembly and assessment (Wigley *et al.*, 2019), and a comprehensive analysis of his work in economic and industrial geology (Eyles, 1969a; Torrens, 2001, 2003) and drainage (Torrens, 2004a). Eyles (1969b) lists Smith's published works, fifty-one in all, that showed an eye-watering wide range of subject matter: geological mapping, drainage works, canal surveying, agricultural methods and improvements, stratigraphy, and palaeontology.

In 1801, William Smith produced a manuscript map showing the distribution of strata in England and Wales, and although the outline of the north-east coast of Ireland is shown on the map, the area inland is not coloured geologically (Henry, 2014, p. 18). This map is now in the collections of the Geological Society of London. Smith's 1815 geological map depicts the eastern margin of Ireland, but it is completely devoid of geological information. Similarly, the north-west portion of France is shown and on it the only geological information provided is that of a thin wash of green colour between Calais and Cap Gris-Nez that allowed for correlation of French strata across the Straits of Dover with the Chalk exposed along the Kent coast. However, five years after the publication of the 1815 map, Smith and his publisher Cary issued *A new geological map of England and* Wales, with

DOI: 10.4324/9781003475552-9

the inland navigations, exhibiting the districts of coal and other sites of mineral tonnage at the smaller scale of 15 miles to the inch. This map provided a representation of Irish geology along its eastern coast from the Giant's Causeway in the north to County Wexford in the south.

What is the source of the Irish geological information shown on the 1820 map? Is there any evidence that Smith was responsible for determining the extent and boundaries of the lithologic units, or was this information acquired by him from other sources? Eyles and Eyles (1938, p. 208) provided one potential answer: 'additions appear to have been taken from a geological map of parts of Ireland and Scotland published in 1816', whereas Henry (2014, p. 24) stated that the sources of this geological information were unknown. On examination of Berger's map of 1816 referred to above, it is immediately clear that this cannot be the whole story, and that Smith did not rely on that source of information alone for his Irish geology.

The sources of information available to Smith are outlined, and through comparison of these with the 1820 map aims to determine those sources, if any, that he used.

William Smith in Ireland

During the 1835, British Association for the Advancement of Science meeting William Smith was honoured through the conferral of a Doctorate of Laws which 'amused and pleased' him (Phillips, 1844, p. 120). The doctoral certificate is now in the Smith Collection, Oxford University Museum of Natural History (ISAD(G) WS/A/3). This significant event took place on 14 August in Trinity College, Dublin (Herries Davies, 1969a). Following the conclusion of the scientific meeting, Smith travelled around the north of Ireland in the company of Samuel John M'Clean, a Fellow of Trinity College. During those travels, they met up with the geologist Joseph Ellison Portlock in Belfast and discussed matters including drainage and agriculture in Ireland (Phillips, 1844, pp. 120–121). This was a subject about which Smith had entered into correspondence with his patron the Duke of Bedford in 1802; he had been requested by some Irish gentry and landowners to undertake surveys and assessments of how best to drain large tracts of the Irish countryside. Smith, who had already been engaged in such work in England from around 1796 (Torrens, 2004a), quoted a price for his services, but apparently, he was never engaged in such schemes in Ireland. There are three letters in this exchange: 25 September 1802, Duke of Bedford to Smith requesting Smith consider the offer of the Gentlemen of Ireland (ISAD(G) WS/C/1/0/015); 14 October 1802, Smith to Bedford, containing his charges (ISAD(G) WS/C/2/0/023); 28 October 1802, Smith to Benjamin Richardson, noting that he was waiting for a reply from the Duke of Bedford to his letter (ISAD(G) WS/C/2/0/026) (see http://www.williamsmith online.com; Oxford University Museum of Natural History). Smith's 1835 Irish sojourn came fifteen years after the publication of his small-scale map, and there is no evidence yet come to light that suggests Smith ever visited Ireland prior to this visit, and so it should be concluded that the geology depicted on the 1820 map was acquired from published or manuscript sources rather than from his own personal field observations.

Ireland and its geology as depicted on Smith's map of 1820

The area of Ireland depicted on Smith's 1820 map extends from the northern coast of County Antrim and Rathlin Island to twenty-two miles south of the town of Gorey, County Wexford, and to the west as far as a line drawn north-south just beyond the western edge of the Giants Causeway and the towns of Randalstown, County Antrim and Ratoath, County Meath (Figure 6.1A). The names used for a number of towns have now fallen out of use: Newborough in County Wexford was so named on a seventeenth-century charter but never found favour over its older and current name, Gorey, while Clarisford in County Wicklow has now disappeared.

Five lithologic units from Smith's 1820 map portray the geology of Ireland:

> n: Coal Districts *with the Rocks and Clays accompanying the Coal comprising the great Iron-works* (coloured Black)
>
> o: Mountain Limestone, *alternating with beds of Shale Chert & Grit-stone and containing numerous Veins of Lead and Copper* (coloured Dark Royal Blue)
>
> p: Trap Rocks (coloured Moss Green)
>
> r: Killas or Slate, *and other Mountain Strata abounding with Minerals* (coloured Blue-Grey)
>
> s: Granite, Sicnite [*sic*] &c (coloured Red)

Smith also provided a table listing rivers and canals on which various mineral resources were transported, but no Irish waterways were given.

Sources of Irish geological information and comparison with Smith (1820)

Geological maps of eastern Ireland

Prior to 1820, there were relatively few publications containing geological information on Ireland (see Wyse Jackson, 2001b for listing) and many of these were the result of county surveys conducted through the auspices of the Physico-Historical Society of Ireland or the Dublin Society (Herries Davies, 1979. However, the geological detail often was scant and for most counties geological maps were not provided. In the context of the eastern portion of Ireland illustrated by Smith in 1820, only Robert Fraser's Wicklow treatise of 1801 contained a map, albeit rather simplified. This map showed five lithologic units—only marked where Fraser considered rock to occur, and described as being unsatisfactory,

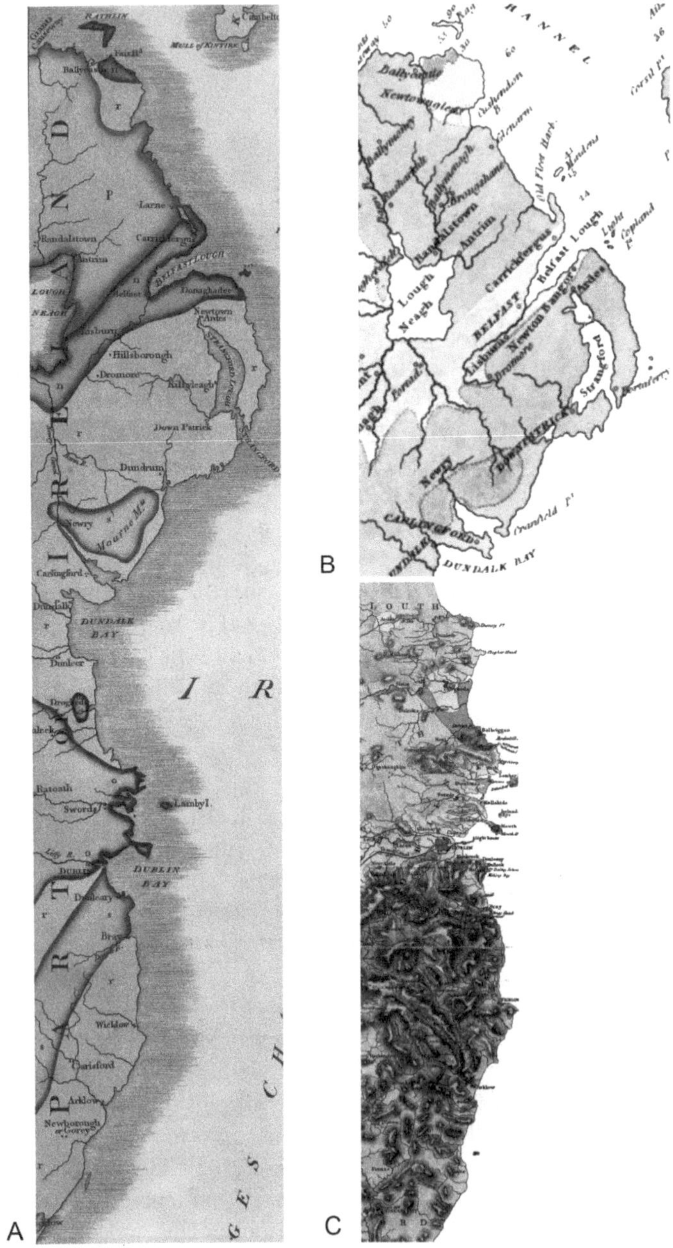

Figure 6.1 Geological map of Ireland by Smith and his map sources. (A) Geology of Ireland as depicted in William Smith's 1820 map; (B) Portion of Berger and Conybeare's 1816 small-scale regional geological map of north-east Ireland; (C) Portion of Weaver's 1819 map showing the geology of south-east Ireland (From Wyse Jackson, 2016).

resembling 'plump, fuzzy, and brightly coloured caterpillars . . . slowly crawling their way across the sheet' (Herries Davies, 1983).

The Dublin Society also promoted geological work through their employment of Richard John Griffith to survey the Leinster coalfield from 1809 and as Mining Engineer from 1812; he generated a detailed report on the coal-bearing districts of the area (Griffith, 1814) and elsewhere in Ireland. Some geological reports and maps were produced by individuals associated with Trinity College and the University of Dublin such as the report on the geology of Dublin and Wicklow by Water Stephens and William Henry Fitton (1812) printed by William Phillips, a founding member of the Geological Society of London. This map shows the north-eastern portion of the Leinster granite on which is accurately marked its southern coastal boundary at Killiney, County Dublin. The map also shows the distribution of the quartz rocks (= Cambrian Quartzite) of the Howth Peninsula, County Dublin, and Bray Head, County Wicklow.

In 1807, the Geological Society of London was established and quickly drew members from Ireland. With the initiation of the *Transactions*, geological information on various parts of the British Isles began to be rapidly disseminated. Through the successful efforts of the Secretary, James Laird, and William Babington in raising funds, the Society was able to engage the Swiss geologist Jean Francois Berger to undertake two seasons of geological mapping in north-east Ireland in 1811 and 1812 (Woodward, 1907). The results of his survey were read to the Society in April 1814 and edited and augmented by William Daniel Conybeare, who had visited the district in 1813 in the company of William Buckland. Berger's report was illustrated with two geological maps and published in 1816 (for portions of these maps, see Figures 6.1B and 6.2).

Further south the Gloucester-born mining geologist Thomas Weaver, who was mine manager at the copper mines at Avoca in County Wicklow for many years (McArdle, 2018), had been assembling a memoir and map on the geological features in the south-east. Weaver had studied in Freiberg under Werner and while in Ireland joined the Dublin Society and was unsuccessful in his attempt to be engaged as either their Mining Engineer or Professor of Mineralogy, losing out to Richard Griffith and Charles Lewis Giesecke, respectively. Weaver later worked in Mexico and died in London in 1855. He published two important memoirs—on the east of Ireland (1819), a paper read to the Geological Society on 15th May 1818, and the south of Ireland (1837) in the society's *Transactions*. The maps that accompanied both are reasonably detailed, although the dating of some of the successions portrayed in the later maps proved problematic. For example, the rocks lining the valleys of Cork and Kerry were said to be Transitional Series but were later assigned to the Lower Carboniferous. A portion of his 1819 map of the east of Ireland appears in Figure 6.1C.

At the same time, Griffith was working towards producing a geological map of the whole of Ireland and this was finally published on a small scale of approximately one inch to ten miles in 1838 (Griffith, 1838). Griffith's cartographic masterpiece at the larger scale of a quarter of an inch to the mile appeared the following year and in various editions until 1855 ([Herries] Davies, 1977a).

It is impossible to know which of these geological publications and maps relating to Ireland William Smith may have had access to. Had Smith relied on the non-Geological Society of London-published sources of Irish geological cartographic information mentioned above, he would have discovered that they provided only a rudimentary outline of the distribution of the different lithologic units. Close examination of his 1820 map and comparison of it with those maps contained in the papers published by the Geological Society of London adequately demonstrates that they were utilised by him in the drawing of geological boundaries for the Irish portion of the map.

While Smith was never admitted as a Fellow of the Geological Society of London, it seems inconceivable that he would not have had access to its *Transactions*. This he gained most likely through his great friend Benjamin Richardson who had been an Honorary Member since 1808 (Herries Davies, 2007). Smith's nephew John Phillips was admitted to Fellowship in 1828 and later served as President (1858–1860).

The two papers and associated geological maps that are of most interest in this Smithian story are those by Berger and Conybeare (1816) and Weaver (1819) published in the *Transactions* by the Geological Society.

Berger and Conybeare (1816)

The small-scale map in Berger and Conybeare (1816, plate 9) illustrates the general geology of north-east Ireland and the south-western portion of Scotland across the North Channel and is rather crude (Figure 6.1B). Five lithologic units are shown: granite (red), mica schist (yellow), greywacke (blue), coal and associated sandstones and limestone (pink), and Floetz trap (modern term = basalt) (green), and as such are the same as used in Smith (1820). Importantly, in this case, the coincidence and general delineation of some of the boundaries with Smith (1820) are more faithful although some differences are obvious. The general positions of the older rocks (the granite and mica schists) on the north-east corner (east of Ballycastle and north of Cushendun Bay) and the south-west-trending boundary between the coal and associated sandstones and limestones (around Belfast) with the Antrim basalts and the position where that boundary meets Lough Neagh are certainly similar. However, the town of Ballycastle is shown underlain by two different lithologies: Floetz trap in Berger and Conybeare (1816) and Killas or Slate in Smith (1820). The Larne Peninsula, which encloses Old Fleet Harbour (now Larne Harbour), is depicted as being of trap in the older map and coal measures in Smith (1820). The granite batholith of the Mourne Mountains is depicted east of Newry, and also to the south, extending onto the Carlingford Peninsula on the older map (Figure 6.1B) whereas in Smith, it is absent from the latter district (Figure 6.1A).

The larger scale map of north-east Ireland provided in Berger and Conybeare (1816, plate 8) (Figure 6.2) produced in association with William Buckland (Herries Davies 1983, p. 29) shows (1) Floetz trap (= basalt) (pale green), overlying chalk (purple), and red sandstone (pink) in County Antrim near Ballycastle; (2) a band of red sandstone, mountain limestone, and coal (pink) running south-west

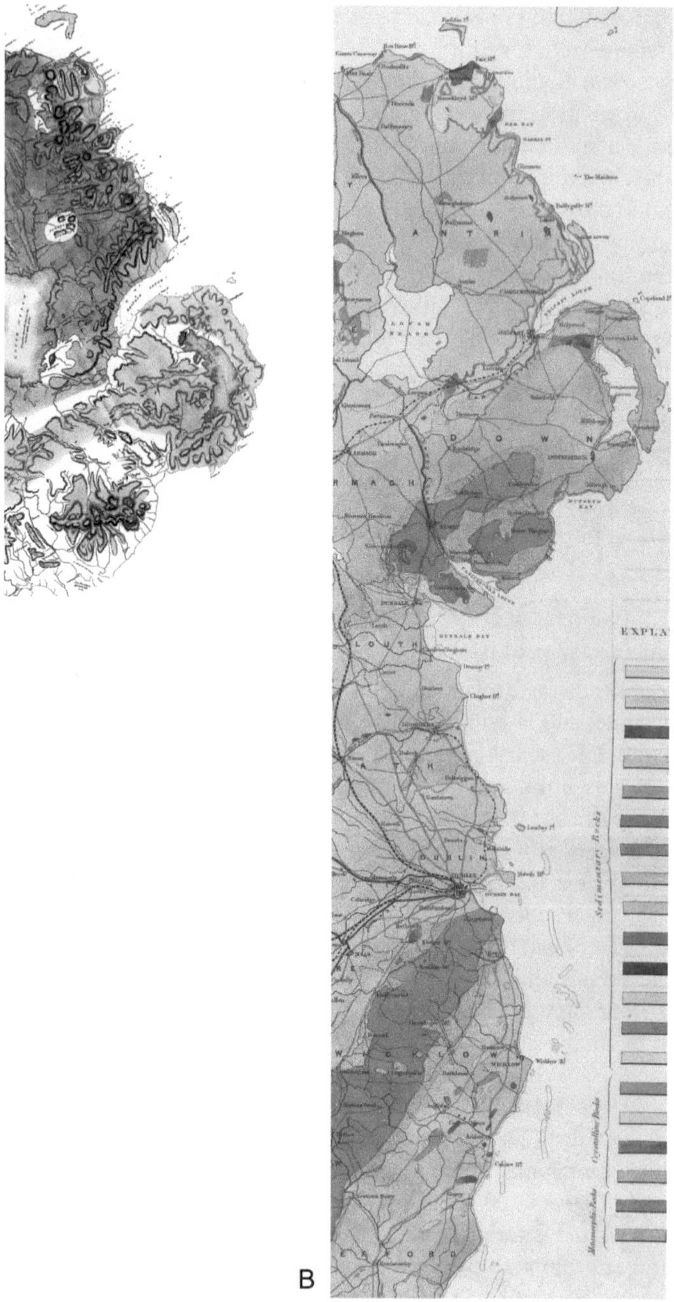

A

B

Figure 6.2 Geological maps of Ireland not utilised by Smith. (A) Portion of Berger and Conybeare's 1816 large-scale map of north-east Ireland; (B) Eastern portion of Richard Griffith's map of 1838 (From Wyse Jackson, 2016).

from Belfast Lough; (3) greywackes (grey) south of Belfast; and (4) a dissected dome of granite (light brown) in the Mourne Mountains to the south. However, when compared with Smith's 1820 map, the boundaries don't coincide. For example, there is a small patch of Antrim basalt that occurs at the south end of Lough Neagh on the 1816 map but is missing from Smith's 1820 map. Likewise, on Smith's map, the Larne Peninsula is assigned to the coal-bearing unit, whereas it is shown as part of the Floetz trap on the 1816 map. Furthermore, the position of the southern boundary of the Coal and Sandstone unit differs on the two maps where that boundary crosses the northern Ards Peninsula. Finally, the extent and degree of dissection of the Mourne granite batholith differs on the two maps. This indicates that the geology depicted on the larger scale 1816 map by Berger and Conybeare did not inform Smith's 1820 map.

Weaver (1819)

Thomas Weaver's 1819 geological map of the eastern part of Ireland covers an area that extends from County Louth on the north to Carnsore Point at the south-easternmost tip of the country (a portion of that map appears in Figure 6.1C). To the west, the map extends from Leitrim to Limerick. Nine lithologic units are shown. North of Dublin, the green-grey 'Clay Slate' grades into 'Transition Rocks' that extend in a tongue to Balbriggan and these surround an outlier of limestone (Mississippian, Lower Carboniferous). This distinctive distribution is replicated in Smith (1820). The wedge of 'Clay Slate' along the north-west margin of the Leinster granite south of Dublin is depicted as pinching out inland at Rathfarnham, County Dublin in Weaver, but it reaches the coast in Smith (1820). The northern margin of the Leinster granite is similar in both maps. The trend of the southern boundary of this igneous body is similar with a south-eastwards bulge of the more southerly part; additionally, the southern extent of this batholith shown in Weaver's map reaches the town of Bray, County Wicklow and Smith's depiction is likewise. In actuality, the true boundary is 7.7 kilometres further north at Killiney, as shown in the earlier cartographic work by Stephens and Fitton (1812).

Weaver provided greater lithological differentiation than later depicted in Smith (1820). For example, surrounding the Leinster granite the mica schist metamorphic aureole is picked out in dark blue; the quartz rocks of Bray Head and further south are clearly marked by the use of a bright yellow colour—Smith did not utilise this division as it would have required him to add another lithologic unit to his key and he had not distinguished this material in Britain. Smith showed the older rocks of the south-east of Ireland as 'Killas or Slate', whereas Weaver subdivided the area into the bipartite 'Clay Slate' and 'Transition Rocks'.

Weaver and Griffith

Richard Griffith, the noted Irish geologist and public servant, writing in a letter to Charles Hamilton, President of the Geological Society of Dublin, implied that Thomas Weaver may have incorporated details from Griffith's work in the 1819 map, remarking caustically that although Weaver's map was different to his as 'to the outline and to the relative position of the rocks' his (Griffith's), he knew 'is correct' and that 'it cannot be said that Mr. Weaver either originated or completed the Geological Map of the east of Ireland' (Griffith in Hamilton, 1846, p. 176). Griffith and Weaver had history: both had been candidates for the position of Mining Engineer to the Dublin Society, Griffith being appointed in 1812. Since 1810, Griffith claimed to be working on the production of a geological map of Ireland and in his 1814 and 1816 lectures to the Dublin Society exhibited parts of this manuscript work. Thomas Weaver had attended these lectures, and so it is entirely possible that some of Griffith's work may have been noted by the mine manager and so found its way into the published pages of the Geological Society's *Transactions* in 1819. However, a comparison of the earliest available Griffith map, dated on the map 1837, but published in 1838 in an *Atlas for the Railway Commissioners*, shows details between the two maps (compare Figures 6.1C and 6.2B) to be rather different, in particular near Balbriggan where, in Griffith's map, the Carboniferous strata reach the coast and an outlier is not depicted. Unfortunately, Griffith's early manuscript maps are not extant (Herries Davies, 1977a, 1983), and so the evolution of his ideas from the early versions of his map in 1816 to the published version cannot be discerned. It is certainly possible that Weaver drew on some of Griffith's ideas and observations, and if this was so, then some of Smith's Irish geological information could be attributed to Griffith's research. This conjecture, however, cannot be confirmed.

Conclusions

It would appear that the northern portion of the geology of Ireland depicted on Smith's 1820 geological map was derived from the small-scale map published as Plate 9 in Berger and Conybeare (1816) as stated by Eyles and Eyles (1938), but that little care was given to transcribing the boundaries accurately from the earlier work.

Given the similarities of most geological boundaries depicted in Weaver's 1819 map with those in Smith (1820), it is probable that William Smith consulted Weaver's publication when drawing the lines for the geology shown on his map. Particularly, striking is the identical outcrop pattern in both maps of the rocks around Balbriggan just north of Dublin where the older rocks, now known to be Ordovician, are shown to surround Carboniferous limestone. Also, the matching margin of the Leinster granite batholith suggests that Smith consulted Weaver's 1819 map.

While Smith did not make any field observations of his own of Ireland's geology until 1835, when drawing the geological boundaries in Ireland for his 1820 map, he utilised the most accurate maps available at the time. Those maps had been published a few years earlier by the Geological Society of London.

7

ALEXANDER NIMMO'S *ON THE APPLICATION OF THE SCIENCE OF GEOLOGY TO THE PURPOSES OF PRACTICAL NAVIGATION* (1825): THE FIRST INVESTIGATION OF MARINE GEOLOGY AND ITS BEARING ON THE GEOLOGY OF OFFSHORE SOUTHERN IRELAND AND SOUTH-WEST BRITAIN

Earth Sciences History 15(2) (1996), 167–171.

Introduction

On 27th October 1823, Alexander Nimmo, Scottish engineer, read a paper entitled *On the Application of the Science of Geology to the Purposes of Practical Navigation* at a meeting of the Royal Irish Academy. This short communication which was published in the Academy's *Transactions* two years later, and in an abridged version in the *Dublin Philosophical Journal and Scientific Review* in March 1825 (Nimmo, 1825a, 1825b), examined the geology of the continental shelf on the south and south-western margin of Ireland and Britain. It is important as it was the first marine geological survey of its kind. It remained for over many years the only contribution to the geology of what is now known as the Western Approaches, the Celtic Sea, the Fastnet Basin, and the Central English Channel. Subsequent biological dredging expeditions from the 1890s onwards threw more light on the geology of the area, but it wasn't until the late 1930s when geological exploration began in earnest, with seismic surveys and drilling operations (Glennie *et al.*, 1987), that a clear picture of its geological structure and stratigraphy was gathered. This area of the continental shelf has now been well traversed by geologists and geophysicists in the exploration of oil and gas, and its geological structure is today very well known and understood (Naylor and Shannon, 1982; Evans, 1990).

DOI: 10.4324/9781003475552-10

Alexander Nimmo

Alexander Nimmo (1783–1832) (Figure 7.1) was born in Kirkcaldy, Fifeshire, Scotland, the son of a watchmaker. He was educated at the Universities of St Andrews and Edinburgh and, after working as a teacher for a number of years, became a Civil Engineer having trained under Thomas Telford. One of his first jobs was to conduct a survey to define Scottish county boundaries (Stronach, 1909). Soon afterwards in 1811, he moved to Ireland where he spent much of his working life engaged in public works. He surveyed boglands for the Irish Bog Commissioners, worked as an engineer for the Office of Public Works, prepared navigation charts of Ireland's coasts, designed over thirty harbours, and surveyed many others for the Irish Fishery Board (Nimmo, 1831). Parts of the docks in Galway City, on the west coast of Ireland, are still known as Nimmo's Pier. He built roads and bridges including a fine single-span bridge of sixty-five feet over

Figure 7.1 Alexander Nimmo. Marble bust in the collection of the Royal Dublin Society (Author's photograph).

a waterfall at Poulaphuca, County Dublin erected in 1828 and another crossing the River Shannon at Limerick constructed seven years later (Wilkins, 2016). He built the village of Roundstone in County Galway in 1822 and settled Scottish fishermen there and did much to enhance Co. Galway (Villiers-Tuthill, 2006). He was also the engineer for various railway lines in England including the Liverpool and Leeds Railway. The results of much of this public work were documented in a number of official reports to various government bodies.

It is not unusual that as an engineer and surveyor Nimmo should have fostered an interest in the developing science of geology as he would have seen excellent geological exposures whilst engaged in surveying various bogs and railway cuttings and embankments. The rapid development of the rail network in Britain and Ireland from the 1830s onwards led to a greatly increased knowledge of the geology of those islands. Indeed, the first large-scale geological map of Ireland was published by Sir Richard Griffith in an atlas prepared for the Railway Commissioners of Ireland in 1839 ([Herries] Davies, 1977a, 1983).

Nimmo joined the Irish Bog Commissioners as a surveyor in 1811 and for much of the next decade traversed much of the western part of Ireland, during which time he made geological observations in several Western counties. An account of this work was published in 1814 in the reports of the Commissioners (Nimmo, 1814).

A map of Ireland, which belonged to Nimmo, and which is now in the National Library of Ireland, carries a crude representation of the geology of Ireland, as it was known at that time. This map dates from 1832 at the latest, and it is possible that the geology marked on the map owes it origin not to Nimmo, but to the efforts of Richard Griffith who was likewise employed by the Irish Bog Commissioners in the early 1800s ([Herries] Davies, 1973).

For a man who was well versed in modern languages as well as aspects of science, it is surprising that only one paper (the subject of this communication) bearing Nimmo's name appears in the *Royal Society's Catalogue of Scientific Papers* of 1870. However, much of his written legacy is to be found in various official reports on public works for Parliament. He also contributed articles to *Brewsters Cyclopedaeia* on Inland Navigation, Bridges (co-authored with Thomas Telford his mentor and tutor), and Carpentry (co-authored with Peter Nicholson).

Apart from engineering, his interests included geology, astronomy, and chemistry, and he was a mathematician of some note. A measure of his standing is reflected in his being elected a member of the Royal Society and the Royal Irish Academy. Nimmo died in Dublin in 1832.

Nimmo's thesis on the geology of offshore south-east Ireland and Britain

Nimmo's paper contains observations made by him whilst he surveyed the coasts, harbours, and fishing grounds of Ireland. He did this in an attempt to gather information which might be used to improve the fisheries of that island. This survey necessitated the taking of soundings from which information about water depth and the nature of the sediment and the geology of the seabed was gathered. Nimmo's survey was the first to gather such geological data and predates Robert Austen's and Henry Marc Brunel's survey of the geology of the English Channel by 25 and 43 years, respectively. Austen described the sediment of the English Channel in terms of its size and shell content and indicated only where granite occurred (Austen, 1850) whereas Brunel who used a gravity corer for gathering geological samples for the first time had been thought to have carried out the earliest marine geological survey (Donovan, 1967).

Nimmo appreciated that coarse clastic material would be found closer to the shore and that muddy sediments were more likely to be found in deeper water. Rough ground, shingle, and stones lay off the east coast of Ireland, beyond which were 'clean ground' and sand, and in the deepest water lay mud and oozes. As the Isle of Man was approached, the muds were replaced by a sandy zone which in turn was followed by coarse clastic sediments. This middle sandy zone was most suitable for trawling for fish and the deeper muddy zone was not, and so there cod was caught on lines and not by trawl.

He noted that in general sailors could detect the type of bottom sediment from the nature of the sounding returned from it and that while charts of these soundings and the nature of the sediments were available at that time, they were not reliable as the accompanying descriptions were vague and not consistent, because:

> … the pilot has never yet been taught to consider the chemical or mineralogical character of the substances, which he brings from the bottom of the sea, so that it is not easy to conjecture what is meant by the descriptions he gives
>
> (Nimmo, 1825a, p. 47)

Nimmo examined the dredgings that he made and subdivided them into five petrological groupings. Approaching the Ireland and Britain from the south-west an area of quartz and granite sand marks the edge of the continental shelf at a depth of 100 fathoms. Similar material is found off Dartmoor. Brown sand and greywacke were found in two areas, south-west of Ireland and off the tip of Cornwall. Due south of Ireland, a large area of mud prevailed that extended from the one-hundred-fathom line east to the Scilly Isles. Lying eastwards of the granite sand towards the English Channel Nimmo found patches of red sandstone conglomerate (probably Devonian Old Red Sandstone) and a broad swath of calcareous shelly sediment. The deepest portion of the English Channel was underlain with this shelly calcareous material.

Accompanying the paper is a map of the geology of the sea south of Ireland and the English Channel and it includes the land areas south of a line between Limerick and Wexford (Figure 7.2). The map measures 18.5 by 9.5 inches (46.9 by 24.2 cm) and was engraved by Jeremy Taylor of North Strand, who may have

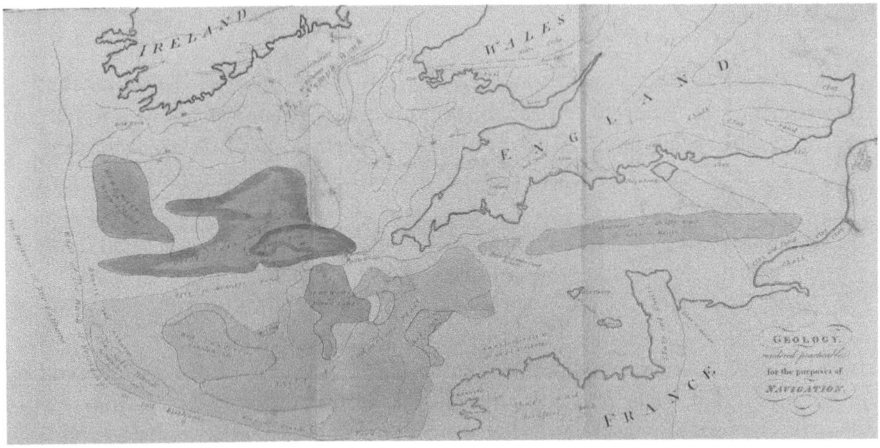

Figure 7.2 Alexander Nimmo's (1825) map of the geology of offshore southern Ireland titled *Geology rendered practicable for the purposes of Navigation* (Geological Museum, Trinity College Dublin).

been one of the Messrs Taylor employed as a surveyor by the Post Office in the 1820s (Nimmo, 1824; Andrews, 1969).

On Nimmo's map which is titled *Geology Rendered Practicable for the Purposes of Navigation*, the geology of the seabed is represented by five tints applied by hand in watercolour: brown = 'Greywacke or Brown Sand'; black = 'Great Ooze Bed'; grey = 'Grey or Granite Sand' or 'Grey and White Sand Granite'; Red = 'Red Sandstone Conglomerate' or 'Red Sand'; Light Blue = 'Calcareous or Shelly Tract, Qu[artz?]. Lias and Oolite'.

The map also includes crude representations of the geology of southern Ireland, south Wales and England, and north-west France. In southern Ireland, although not labelled as such, the northern boundary of the Devonian rocks is marked and the Carboniferous-floored eastwest trending valleys are also delineated. The geology of Wales is represented as slate rocks to the north of an area of coal along the south coast. In England, the granite mass of Dartmoor in Cornwall is marked as well as chalk and various areas of clay, which extend across the English Channel to France. The rocks of Brittany are divided into 'Slate and Primative Rock' and 'Granite'.

Nimmo contended that:

> Nothing is more interesting to navigators, on their way home from the Atlantic to our coasts, than the means of distinctly ascertaining their position on approaching the land. In thick weather, when astronomical observations are not to be had, and when, from the influence of unknown currents or the like, the reckoning may be justly suspected, it is of the utmost consequence to keep the ship in the fair way of the channel
>
> (Nimmo, 1825a, p. 45)

He added that:

> It appears to be therefore possible, on rational principles, to delineate masses of the different soils of the bottom of the sea, as has been done for those of the land; and, by instructing the pilot more carefully to ascertain the distinctions between the various kinds of sand or other matter which he may bring up from the bottom, so as more readily to refer to the corresponding part of his chart, it is evident, that a great additional precision will be afforded to this mode of discovering his place
>
> (Nimmo, 1825a, p. 49)

Unfortunately, it is not known how this novel aid to navigation was received by sailors or by geologists, and although Nimmo hoped further observations would refine the map, it appears that no improvements subsequently appeared. Nimmo had hoped to produce similar maps of other areas, but these were never published. It is possible that lying in some drawer in an archive, long forgotten, are manuscript charts in Nimmo's hand that depict the geology of other parts of the continental shelf around Ireland and Britain.

Later studies of the geology of offshore Ireland and south-west Britain

Nimmo's paper of 1825 remained for many years the only treatment of the geology of offshore southern Ireland and south-west Britain. However, it is not well-known and citation of it has not been found in any subsequent publications dealing with the geology of the area. Since the work of Jean-Étienne Guettard (1715–1786), geologists have recognised similarities between the geology of southern Britain and northern France (Guettard, 1746; Prestwich, 1855–1857) and between Ireland and western Britain (Jukes, 1866; Hull, 1880). However, these connections were purely based on observations made on dry land, and attempts to delimit strata beneath the stretches of water separating the various areas were not made until the 1920s. It was then that dredging by French scientists showed that the geology of the English Channel was not continuous with that exposed either side of it (Dangeard, 1929; King, 1955).

Just over forty years after Nimmo's work, Brunel in 1866 carried out a geological survey of the English Channel between Dover and Calais. This survey area lay just east of Nimmo's. At the time there was considerable interest in driving a tunnel between south-east England and France—an engineering feat completed 128 years later. The work of Brunel, together with William Topley (1841–1894) who in 1872 drew a cross-section of the geology across the narrow strait (Topley, 1872), and later French-dredging expeditions of the 1920s resulted in the marine geology of this part of the sea around Britain being well-understood by the 1930s. Such was not the case with the Western Approaches and the Celtic Sea off southern Ireland.

Further information on the geological structure of the latter was gathered almost incidentally with the commencement of oceanographic expeditions such as those of the 1860s and 1870s which climaxed in that of the H.M.S. *Challenger* (1872–1876) and continued into the first decade of the present century. These primarily gathered biological information, but other data in various disciplines including geology was compiled. From the *Challenger* expedition, over 12,000 samples were described (Murray and Renard, 1891). Dredging expeditions were particularly popular at this time and between 1895 and 1906, a series of dredging expeditions organised by the Marine Biological Association out of Plymouth gathered data on the geology of the English Channel (Crayshaw, 1908; Worth, 1908). These showed that crystalline rocks similar to those cropping out on mainland Britain were to be found close to the shore, while Mesozoic and Tertiary rocks occurred in deeper waters. In the early 1900s, material from the Porcupine Bank 130 miles west of the Irish coast was recovered by the Department of Agriculture and Technical Instruction for Ireland. This survey recovered a number of specimens which demonstrated the presence of Mesozoic and Tertiary strata offshore western Ireland (Cole and Crook, 1910).

The nature of the geological successions in the area covered by Nimmo was fully elucidated over forty years later when exploration into the geological nature of the area around Britain and Ireland started seriously in the 1940s. Seismic studies, gravity and magnetic surveys, seabed sampling, and palaeontological research indicated the positions of various offshore basins which were filled with post-Palaeozoic sediments. The earliest seismic studies were carried out in 1938 and 1939 (Bullard and Gaskell, 1841) and continued by British and French scientists, working for government agencies, commercial interests, or university departments particularly at Bristol under W.F. Whittard (Curry *et al.*, 1970).

After the discovery of gas in Holland, the basins of the North Sea area became the focus of fervent exploration, from the 1960s onwards, in the hunt for further hydrocarbon reserves. Today many are producing oil and gas in considerable volume. The success of the North Sea exploration and drilling led to the exploration companies turning their attention to the continental shelf around Ireland and south-west Britain. Exploration focussed on the Celtic Sea Basins and the Fastnet Basin to the south of Ireland and in the Western Approaches Trough, areas that lie within the limits of Nimmo's survey. In 1971, a large gas field was discovered 30 miles south of Cork (Collins, 1976), which for several decades met most of Ireland's domestic gas demand, and in 1974, the first well in the British sector was drilled (Evans, 1990). Exploration and research have declined in the last two decades, but most of Ireland's and Britain's offshore basins have been drilled and their geological structure described and understood as fully as possible.

Conclusion

Nimmo's paper and map of 1825 represent an important milestone in the history of geology in Ireland and Britain. It was the first attempt at offshore geological mapping, although the reasoning for producing such a map—for aiding navigation—does not appear to have been thoroughly successful. Little did Alexander Nimmo realise that his brief foray into offshore geology would be followed nearly 120 years later by extensive and expensive exploration of all areas of the continental shelf of western Europe. Nimmo deserves to be remembered by historians of geology, if only for his pioneering thesis of 1825.

8

JOHN W. PRINGLE (*C.* 1790–1861) AND ORDNANCE SURVEY GEOLOGICAL MAPPING IN IRELAND

Proceedings of the Geologists' Association 108 (1997), 153–156.

Introduction

In a paper tracing the relationship between the army and some of the founding fathers of British geology, Rose (1996) cited as an example the early development of government geological mapping in Ireland by Royal Engineer officers. He noted that the work was begun under Trigonometrical (Ordnance) Survey auspices by Captain John W. Pringle on the initiative of Lieutenant-Colonel Thomas F. Colby; that it was re-activated later by Captain Joseph Ellison Portlock, and that when the Geological Survey of Ireland was founded in 1845 as a civilian organisation its first Local Director was (briefly) Captain H. James. The work of Portlock and James is relatively well known, but that of Pringle is not. Publication of an official history of the Geological Survey of Ireland (Herries Davies, 1995) that extends an earlier account (Herries Davies, 1983) and Rose's two papers (1996, 1999) together with this contribution add more documentation and assessment of Pringle's role in the ancestry of the Irish survey. It is as John Watson Pringle (Herries Davies, 1983, p. 88; 1995, p. 8) rather than John William Pringle that he was known in Ireland, and in *The Army List* from at least 1843 to 1860, Rose (1999) adds much to what is known about Pringle's early and personal life and argues that he was born in either 1790 or 1791 rather than in 1973 as given by me in the original of this paper; additionally, he noted that after about 1836, Pringle had replaced 'William' with 'Watson'.

J.W. Pringle and his work in Ireland

In 1825, the Ordnance Survey began topographic mapping in Ireland at the scale of six inches to the mile under the direction of Thomas Colby. In order to advance his interest in geology, Colby directed his surveyors to collect rock samples whilst

 DOI: 10.4324/9781003475552-11

mapping and to record on maps the location from where they had been collected. These specimens were then to be sent to Dublin for examination. The ultimate aim of this work, which was secondary to the main task of topographic mapping, was to produce geological maps.

This section of the Ordnance Survey was to be headed by a 'Superintendent of the Geological Survey of Ireland', a position offered by Colby to, and accepted by, John W. Pringle (*c.* 1793–1861), who started geological duties on 14 November 1826. Pringle, who had studied under Friedrich Mohs at the Freiberg School of Mines (Portlock, 1869), and who was highly regarded as a mineralogist, had been working in Ireland with the Ordnance Survey from June of that year.

For over a year, he was continually engaged in baseline surveys and was unable to pay much attention to geology. In December 1827, he visited surveyors in the basaltic region of north-east County Antrim, and there he discovered that the men knew little geology and collected specimens in a way that proved them to be useless for the purpose for which they had been acquired. He ordered his men to recollect material from the poorly collected areas, a task that some of them openly questioned, reasoning that determining the underlying geology was a waste of time. In due course, Colby asked Pringle to cover the ground himself.

Soon after this event, Pringle published a small seventy-page instruction manual, *Directions for geological and mineralogical observations* (Pringle, 1827), that outlined the methods that should be used by the surveyors in collecting geological information. The volume was initially handwritten (not by Pringle) and from this master copy, fifty copies were lithographed at the Tower of London and subsequently distributed to each surveyor in Ireland. Pringle gave detailed information on the required size for specimens and on how they should be wrapped, and on what observations (dip and strike of strata etc.) should be collected and recorded on maps. The volume also contains an interesting hand-coloured sketch map of the geology of southern England. The only known copy of Pringle's manual is now in the archives of Geological Survey Ireland. It is signed by him, and liberally annotated by Portlock, a fellow Royal Engineer and geologist.

Other initiatives aimed at enhancing the geological knowledge of the surveyors included the distribution of sets of rocks and minerals, German and Cornish, as well as the purchase of the *Transactions of the Geological Society of London* and some other geological texts for their use. Soon afterwards, Pringle's men were asked to produce, in addition to geological plans of each parish, two geological cross-sections through the area, in a north-south and east-west direction. Two of these crude, and early examples of the genre survive in Dublin ([Herries] Davies, 1974), show sections through parts of County Derry where Tertiary basalts overlie Mesozoic rocks.

Colby's attempt to establish a geological survey was commendable but largely doomed to failure. Not one published geological map emanated from offices of the Ordnance Survey during Pringle's period in charge, although it was his and Colby's intention to publish and sell such maps produced on a parish scale.

Apart from the cross-sections mentioned above, there survives a map of the parish of Aghanloo, County Derry dating from 1828 on which coloured washes have been added that indicate the distribution of various lithologies, and on which the localities from which the samples were collected have been marked.

In July 1828, a fellow Royal Engineer, Sir James Carmichael Smyth, was sent from England charged with the task of examining the workings of Colby's survey. It is coincidental that Carmichael Smyth was Pringle's superior officer at the Battle of Waterloo (Dalton, 1890, 1904). He was particularly concerned that valuable time and indeed money were being spent on a geological survey which had not received official sanction. Pringle stated that geological inquiries were completed for three-quarters of those parishes already surveyed topographically, and that specimens, plans, and sections were in Dublin. He also made the astonishing claim that geological coverage for the whole country could be completed by 1834—a claim made all the more ludicrous in that only five of Pringle's men had by then been trained in geological surveying. One of these was Portlock, whose geological prowess was not in question, while the other four had limited geological experience. Carmichael Smyth rapidly deduced that the geological survey was largely ill-conceived and poorly run and on 1 September 1828 directed that all geological surveying should stop immediately. Pringle's Irish geological sojourn came abruptly to an end after less than two years.

Thomas Colby in 1829 managed to get permission to recommence the work, thus making another attempt at geological surveying; but this enterprise had little success and few tangible results. He appointed Portlock to oversee the geological surveying, and it was he who was responsible for the geological map and narrative that was published by Colby in 1837 as part of the Ordnance Survey's memoir on the parish of Templemore in County Derry. In the same year, Portlock established a geological office in Belfast and engaged a number of civilian geologists as part of his team. The civilians included Thomas Oldham (1816–1878), who later became the second Professor of Geology at Trinity College Dublin succeeding John Phillips (1800–1874), who held the Chair for barely a year (Wyse Jackson, 1994), and subsequently Local Director of the Geological Survey of Ireland and the Superintendent of the Geological Survey of India; also George Victor Du Noyer (1817–1869) a noted artist. Their work culminated in the magnificent *Report on the Geology of the County of Londonderry, and parts of Tyrone and Fermanagh* (Portlock, 1843), but on account of economic stringencies, Portlock's team was disbanded in 1840. Many of the specimens collected by Portlock and his team are now to be found in the British Geological Survey collections at Keyworth, Nottinghamshire, and in various Irish institutions including the Ulster Museum, Trinity College Dublin, University College Galway, and the Geological Survey of Ireland (Tunnicliff, 1980; Harper and Parkes, 1996).

It was not until the establishment of the Geological Survey of Ireland in 1845 that geological mapping was placed on a firm footing and a rational and well-conceived programme commenced. It was to take this survey forty-five years and a huge capital outlay to complete the primary geological mapping of the country—considerably longer than the estimate made by Pringle of a completion date of 1834 and cost of £5,500 (Herries Davies, 1974). It would appear that John W. Pringle underestimated the ability of his surveyors, the difficulty of the Irish terrain, and the depth of the Treasury's pockets during his short but eventful career in Ireland.

9

SIR RICHARD JOHN GRIFFITH (1784–1878): A PORTRAIT IN OILS BY STEPHEN CATTERSON SMITH (1806–1872)

Irish Journal of Earth Sciences 26(2008), 45–51.

Sir Richard John Griffith

Richard John Griffith (1784–1878) was one of the foremost public servants in Victorian Ireland and is best known outside geological circles for his Valuation Survey (Lee, 1980; Vaughan, 1980; Reilly, 2000). To geologists both in Ireland and elsewhere, he is primarily remembered for his large-scale geological map of Ireland published by the Railway Commissioners that was first published in 1839 and which went through various editions up until 1855 (Herries Davies, 1977a).

It is surprising, given his status, that there are few pictorial representations of Griffith. A pastel drawing dated 1840 when he was fifty-six is reproduced as the frontispiece to the volume published in 1980 to mark the centenary of his death (Herries Davies and Mollan, 1980) and also in Herries Davies (1983, p. 37), while a photograph of Griffith in 1874 probably taken to mark his ninetieth birthday was reproduced by Praeger (1949, plate 5, facing p. 88) and by Herries Davies (1983, p. 78). An earlier photograph was donated to the Geological Society of London in 2022 and will be the subject of another communication elsewhere. A bust of Griffith late in life by Sir Thomas Farrell is in the collections of the Royal Dublin Society (reproduced in Herries Davies, 1980, p. 25) while another by Christopher Moore dated 1859 is in the Ulster Museum. A cartoon of him clad in trilobites (several *Griffithides* species) attacking John Phillips, similarly clad (in a number of *Phillipsia* species) also appears in Herries Davies (1983, p. 74).

In 2003, the National Gallery of Ireland acquired an oil portrait of Griffith (Figure 9.1), painted by Stephen Catterson Smith (1806–1872). It is interesting given the array of geological trappings in the composition of the portrait.

 DOI: 10.4324/9781003475552-12

Figure 9.1 Portrait of Richard John Griffith by Stephen Catterson Smith painted in 1854 (© National Gallery of Ireland).

The portrait

The portrait measures 50 by 40 inches (127.6 cm by 102.2 cm) and is a three-quarter-length view of Griffith seated in an ornately carved chair. He is dressed in a black frock coat, white shirt with a black cravat, and grey trousers. The picture is encased in a rather simple gilt frame. The portrait requires some restoration as there is some cracking of the paintwork that is most obvious on the sitter's forehead, and some paint has flaked off along its lower margin.

The painting includes a number of objects, which on close inspection are all of geological interest. On Griffith's lap lies a geological map of Ireland supported on one end by his right hand and on the other is wedged between the tabletop and a book on which his left forearm is resting. In the original publication of this paper, I identified this map as Griffith's small-scale geological map of Ireland printed in colour by Forster and Co. of 2 Crow Street, Dublin and dated August 1853. However, Gordon Herries Davies and Jean Archer recognised the map as the small-scale 1838 water-coloured map that had been published in the *Atlas* to accompany the second Railway Commissioners Report (Anon., 1838); Griffith had been appointed a Commissioner in 1836 and used the position to further his geological cartographic ambitions. On this map, the different lithologies are highlighted in a palate of twenty watercolour swatches, whereas the later 1853 map carried 37 different colours. The older map carries a title as does the 1853 map which also details fossils found in different geological formations. This printed material is not given on the 1838 map and is not seen in Smith's portrait. Further details of the 1853 map printed for the use of Griffith's surveyors employed in the Valuation Office are given elsewhere (Wyse Jackson and Hawley, 2023).

The book on which Griffith's left arm is resting is almost certainly a copy of Frederick M'Coy's *A Synopsis of the Characters of the Carboniferous Limestone Fossils of Ireland* that appeared in 1844. The volume is the correct size for this conclusion. It had been printed in Dublin for Griffith and all copies known to this author are presentation copies from Griffith. M'Coy (*c.* 1823–1899) had been employed by Griffith to identify and describe the fossils collected by the valuation surveyors who had been ordered by Griffith to collect geological material (Wyse Jackson and Monaghan, 1994 [Chapter 10 herein]). This material is now shared between the National Museum of Ireland (Monaghan, 1992) and the Geological Museum of Trinity College Dublin. The bulk of the material which numbered 4,500 specimens, including 500 types, was presented by Griffith to the museum of the Royal Dublin Society, which now makes up the National Museum of Ireland. In 1849, he presented Trinity College Dublin with the residue of the collection that comprised some 1,000 specimens, and the following year the university awarded Griffith an honorary law degree, no doubt to go some way to acknowledge his gift. Unfortunately, in the 1960s, labels were removed from many of Griffith's specimens in Trinity College and today less than fifty specimens have been confirmed to date as being part of his collection.

On the table behind Griffith lies a medal in an open case. This is the Wollaston Medal awarded to Griffith on 17 February 1854 by the Geological Society of London and is the highest honour the society can award. Edward Forbes, the president of the society, in his citation, noted that the medal was awarded 'for the valuable services rendered by you to geological science and particularly for your Geological Map of Ireland, the result of your own laborious and judicious researches' (Forbes, 1854, p. xix).

Normally struck in gold, Griffith's Wollaston Medal was unusual in that it was one of a batch awarded between 1846 and 1866 that were struck in palladium, a rare white metal discovered by William Hyde Wollaston (1766–1828), the chemist who is commemorated by the award. Griffith was one of two Irishmen to be awarded a palladium Wollaston Medal; the other was William Henry Fitton (1780–1861), a former president of the society, who received his in 1852 (Wyse Jackson, 1998a [Chapter 5 herein]). The location of Griffith's medal is unknown.

Additional books are propped up on the table behind Griffith and although it is difficult to identify these with any certainty, it is possible that among them are his reports for the Bog Commissioners or his reports on various Irish coalfields surveyed by him for the Royal Dublin Society. Papers and possibly geological maps can also be seen faintly behind the sitter.

All the evidence indicates that this portrait was painted in 1854 when Griffith was 70 years of age. In 1858, he presented a photograph of the portrait to his brother-in-law, John Waldie, who hung it close to another image of the geologist, a profile in black ink, in his bedroom at Hendersyde Park in Kelso, Scotland (Waldie, 1859, p. 123). According to Stewart, the painting was exhibited in Dublin at the Royal Hibernian Academy exhibition of 1854 (Stewart, 1987, pp. 162–163).

As is indicated by a label on its reverse, the portrait was also exhibited in 1873 at the Dublin Industrial Exhibition Palace and Loan Museum of Art Treasures, and Stewart (1995) notes that it was lent to the institution by Griffith. This makes it clear that he owned the portrait, and it is not fanciful to suggest that the great man commissioned it himself to mark his seventieth birthday. It probably hung at Griffith's home at 2 Fitzwilliam Place, Dublin, until his death in September 1878. It was possibly then moved to his son's residence at Hendersyde Park, where it would have remained until the family moved from the house in the mid-twentieth century.

In 2003, it appeared in pre-sale exhibitions of Irish art due for auction that were held in Belfast (where it was first seen by Dr Peter Crowther of the Ulster Museum) and later in Dublin. At that time, the National Museum of Ireland considered bidding for the painting, but subsequently, it was agreed that the National Gallery of Ireland should try to buy the painting as it would make a valuable addition to their collection of Irish portraits (N. Monaghan pers. comm., November 2008). The painting was purchased in London on 15 May 2003 by the National Gallery of Ireland.

Stephen Catterson Smith

Stephen Catterson Smith was born in Skipton, Yorkshire, in 1806, the son of a coach-builder, and trained as an artist at the Royal Academy and in Paris (Williamson, 1905, p. 95; Cullen, 2004). He moved to Londonderry in 1839 to undertake some commissions, and later to Dublin. He remained in Ireland for the rest of his life and became the country's 'most fashionable portrait painter of his day' (Stewart and Cutten, 1997, p. 428). He painted portraits of seven Lord-Lieutenants and two of Queen Victoria, including one commissioned by Dublin Corporation (Cullen, 2004). He also created many portraits of the men of the day including Sir Philip Crampton, the celebrated surgeon, and William Dargan, the railway magnet. Richard Griffith was in good company.

Stephen Catterson Smith was a regular exhibitioner at the Royal Hibernian Academy (Stewart, 1995) of which he was president on two occasions from 1859 to 1866 and 1868 to 1869. For many years, he lived at 42 St Stephen's Green, where one assumes he had his studio, and to where in 1854 Griffith probably walked the short distance from his own home for several sittings. Catterson Smith collapsed while hanging paintings for the Dublin International Exhibition

(Cullen, 2004), and he died at home on 31 May 1872. He was buried in Mount Jerome Cemetery in Harold's Cross, where Griffith himself was interred nearly six-and-a-half years later.

Significance of the Griffith portrait

This portrait clearly indicates the area of his activity that Griffith felt to be his most important, and which he wished to have captured in perpetuity in oils. The portrait shows a version of his geological map of Ireland, the book in which his geological work was partially described, and the medal which was awarded to him by his peers for this work. There is no doubt that Griffith considered himself to be a geologist of some significance and this portrait confirms his assertion, although we now know that most of the fieldwork that underpinned his geological map was undertaken by Patrick Ganly (Archer, 1980). It is interesting to note that on receiving the Wollaston Medal in 1854, Griffith in response to the citation strongly indicated that he was solely responsible for the delineation of the geological sequences marked on the map (Griffith, 1854). Griffith's self-importance can also be measured in the artist that he commissioned; unsurprisingly, he sought the best portrait artist of the time to undertake the task. Given Catterson Smith's portfolio of influential members of Dublin society, Griffith probably counted himself as equal or greater to those men, in 1858 being created a baronet.

In 1874, when he had reached 90 years of age, Griffith went to the studio of Thomas Cranfield of Grafton Street where he posed for a photograph (Herries Davies, 1983, p. 78) and it is fascinating to note the similarities in the composition of this image with that of Catterson Smith's portrait of twenty years earlier. In the photograph, Griffith is seated on a chair, dressed in similar clothing, and is holding his 1853 geological map on his lap. His left arm is resting on two books, one without doubt a copy of either M'Coy's celebrated monograph of 1844 or the second edition of 1862. That Griffith requested a photograph posed in an almost identical fashion suggests that either his opinion of his status and importance as a geologist had not diminished over time or that his right side provided a more pleasing view than did his left.

Conclusion

There is no doubt that Sir Richard John Griffith was an important public servant in nineteenth-century Ireland, and as such it is not surprising that a large portrait in oils was painted in his lifetime. The painting clearly indicates that Griffith considered his geological studies to be his greatest accomplishment. It is appropriate given his work to further the development and well-being of Ireland—through his work with the Bog Commissioners, the Royal Dublin Society, the Railway Commissioners, the Office of Public Works, and the Valuation Survey—that his portrait is now part of the collection held by the National Gallery of Ireland.

10

FREDERICK M'COY: AN EMINENT VICTORIAN PALAEONTOLOGIST AND HIS SYNOPSES OF IRISH PALAEONTOLOGY OF 1844 AND 1846

Geology Today 10(1994), 231–234 [co-authored with Nigel T. Monaghan].

Frederick M'Coy

To most modern-day students of palaeontology, the name Frederick M'Coy (*c.* 1823–1899) is familiar; to other Earth scientists, it possibly won't be. Who was M'Coy and why was 1994 a significant year for palaeontologists and historians of the Earth sciences? The year marked the sesquicentenary of the publication of M'Coy's *A Synopsis of the Characters of the Carboniferous Limestone Fossils of Ireland* (1844). This monograph, together with that for Lower Palaeozoic fossils, *A Synopsis of the Silurian Fossils of Ireland*, published two years later, which describe the collections of Richard Griffith, remain highly important to palaeontologists working today.

Adrian Rushton (1979) noted that M'Coy's name has frequently been spelt in a number of ways: McCoy, MacCoy, Mac Coy, Mac`Coy, and, most frequently, M'Coy. The man signed himself as M'Coy and so, as Rushton suggests, this form should be used.

Very little is known of Frederick M'Coy's early life. He was born in Ireland, possibly in Dublin in 1817 or 1823 (which date is unclear), into a Roman Catholic family. His father, Simon, was a medic who taught anatomy in a number of Dublin institutions from the 1820s onwards and who in 1849 became Professor of Materia Medica in Queen's College, Galway, a position he held until his death in 1873.

Frederick (Figure 10.1), who was described by Herries Davies in 1983 as having 'a sturdy frame, reddish hair, florid complexion, and a warm heart', showed from an early age an aptitude for natural history and in 1838 delivered a paper 'On the sternal apparatus of birds as applied to their classification, and adapted to their different modes of flight' to the Dublin Natural History Society, of which his

DOI: 10.4324/9781003475552-13

Figure 10.1 Frederick M'Coy (*c.* 1823–1899) in his youth. Lithograph by Frederick Schoenfeld. Printed by Hamel and Co., 49 Queen Street, Melbourne (Courtesy of the State Library Victoria).

father was also a member. In 1841, he was appointed Curator of the Geological Society of Dublin and was responsible for compiling a catalogue of its museum (1841), as well as contributing a number of papers on palaeontological topics to its journal. However, he fell out with the Council of the Society and was replaced as Curator by Thomas Oldham, who was later appointed the first Local Director of the Geological Survey of Ireland.

In 1840, M'Coy began working on the collections of Palaeozoic fossils that belonged to Richard Griffith and published his findings in two volumes in 1844 and 1846. In 1845, he took up a position with the fledgling Geological Survey of Ireland, hoping to be its palaeontologist, but internal Survey politics directed him to fieldwork, at which he was not particularly skilful. The following year he was accused of sloppy fieldwork by Oldham, which led to his resignation. He then sought refuge in Cambridge where for a number of years he worked for Adam Sedgwick, engaged in describing the fossils in the Woodwardian Museum.

His findings were published between 1851 and 1855 in an important volume, co-authored with Sedgwick, entitled *A Synopsis of the Classification of the British Palaeozoic Rocks*. A synthesis of his knowledge of the Palaeozoic palaeontology of the British Isles was published as *Contributions to British Palaeontology* in 1854. It should be remembered that he was at this time, together with John William Salter (1820–1869), one of the leading lights in this developing discipline. In 1849, he returned to Ireland and took up the Chair of Geology and Mineralogy at Queen's College, Belfast. He retained this position until 1854 when he moved to Melbourne, Australia, to fill the Chair of Natural History.

In Australia, he lectured on a large range of subjects in the university, was Founder and Director of the National Museum of Victoria in Melbourne (Sheets-Pyenson, 1988), whose collections were acquired mainly through purchases made by M'Coy from British dealers, and acted as palaeontologist to the Geological Survey of Victoria. Although his public and university duties curtailed his research output, he still found time to produce the widely acclaimed *Prodromus of Palaeontology of Victoria* and *Prodromus of Zoology of Victoria*, as well as many shorter scientific communications (see list in *Geological Magazine* 36 (1899), 285–287). In later life, he was recognised by his peers by the award of the Murchison Medal by the Geological Society of London in 1879, through election to Fellowship of the Royal Society the following year, through the conferring of a D.Sc. from the University of Cambridge, and ultimately by being knighted in 1891. He died in Melbourne on 13 May 1899 and was buried in Brighton General Cemetery just south of the city close to his residence. He was survived by a son, Henry, who practised as a solicitor.

The Griffith Collection

Richard John Griffith (1784–1878) was an important member of the scientific community in Dublin in the 1800s. He had trained as an engineer in Dublin and Edinburgh, where he also attended the geological lectures of Robert Jameson. On his return to Ireland in 1808, he was elected to membership of the Dublin Society and was later appointed as Mining Surveyor to the Dublin Society. He subsequently became a prominent member of the Geological Society of Dublin (later the Royal Geological Society of Ireland), which had been established in 1831.

Griffith's long-standing ambition was to produce a large-scale geological map of Ireland. In 1830, he was appointed Commissioner of the General Survey and Valuation of Rateable Property, which was to survey the value of land throughout Ireland. He accumulated a large collection of Lower Palaeozoic and Carboniferous fossils by charging his many surveyors to make geological observations and to collect geological specimens while carrying out their official duties. This material was essential to allow accurate determination of the geological age of many of the tracts surveyed by Griffith's men. By 1836, he was also a Commissioner for Railways and he persuaded his fellow members of the merit of publishing a large-scale geological map of Ireland. It appeared in May 1839 and went through several editions until 1855 (Herries Davies, 1977a).

Griffith was also keen to publish a record of his fossil collection but was not a palaeontologist himself. In Dublin, at that time there were few persons apart from M'Coy who could have carried out the work, and so between 1840 and 1844, M'Coy was employed to study and describe the contents of Griffith's and others' fossil cabinets, including those of the Royal Dublin Society and the Geological Society of Dublin.

The Griffith collection numbered several thousand specimens, of which most were donated to the museum of the Royal Dublin Society between 1844 and 1872 and can still be found in the collections of the National Museum of Ireland. A large collection of over 1000 specimens was presented to the museum of the University of Dublin in March 1845, and the following year, 500 specimens were donated to the Woodwardian Museum (now the Sedgwick Museum) in Cambridge. Both these collections contained numerous syntypes of M'Coy species. Unfortunately, much of the University of Dublin collection can no longer be recognised in its present Geological Museum, although a 16-page manuscript catalogue of the donation indicates its size, scope, and importance (Monaghan, 2025). The value attached to the donation by the university is illustrated by the award of an honorary LL.D. to Griffith, its donor, in 1849.

M'Coy's monographs of 1844 and 1846

In 1840, M'Coy began to carry out research on Griffith's collection and on a number of other smaller collections. In 1842, Griffith and M'Coy produced and issued privately under the former's authorship, a slim twenty-five-page pamphlet entitled *Notice Respecting the Fossils of the Mountain Limestone of Ireland*, which heralded the two monographs of 1844 and 1846. These were *A Synopsis of the Characters of the Carboniferous Limestone Fossils of Ireland* (1844) [2nd edition, 1862] and *A Synopsis of the Characters of the Silurian Limestone Fossils of Ireland* (1846) [2nd edition, 1862], published at Griffith's expense. Together they contain descriptions of 28 new genera and 522 new species, as well as numerous taxa already known to science. Although Griffith and his men had made large collections of Lower Carboniferous fish and plants, accounts of these were not included. An account of the fish fossils appeared in the *Annals and Magazine of Natural History* in 1848.

The Carboniferous synopsis was published in October 1844. It is a large volume of 207 pages (9.5 by 11.75 inches) containing descriptions of over 1000 invertebrate species, of which 450 were new. Most of these new taxa are illustrated in the twenty-nine lithographs drawn by George Victor Du Noyer (1817–1869) and H.F. Hutchings (Figure 10.2). The taxonomic descriptions are presented in the following order: Mollusca [Cephalopoda, Pteropoda Gastropoda, and Dithyra and Atrachia (bivalves)] Brachiopoda, Crustacea (including trilobites and ostracods), Annelida, Echinodermata, and Zoophytes (bryozoans and corals).

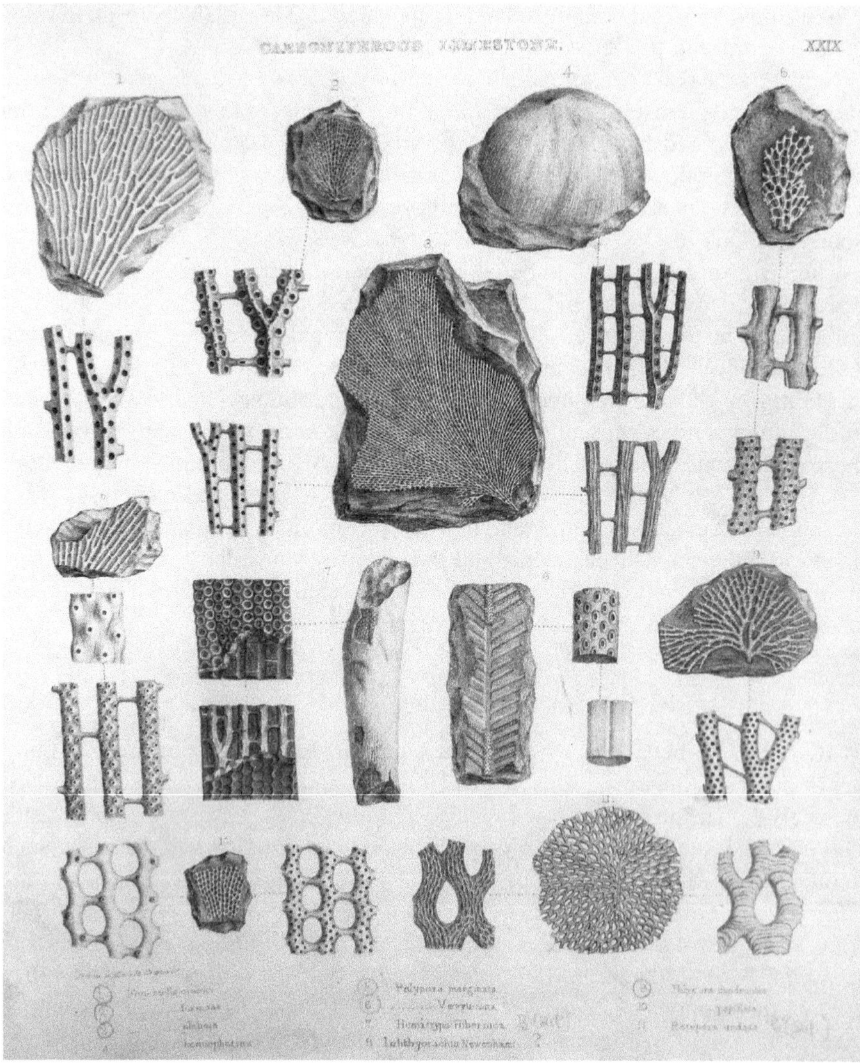

Figure 10.2 Plate 29 from *A Synopsis of the Characters of the Carboniferous Limestone Fossils of Ireland,* illustrating bryozoans. Drawn by H.F. Hutchings (Geological Museum, Trinity College Dublin).

The descriptions are in the main short, being telegraphic in style, with perhaps ten lines devoted to each new species and as little as a solitary line or two for previously described species. Frequently, M'Coy makes no reference to the dimensions of specimens, and scales are lacking in the illustrations. The pictures are reasonably clear, and important features which distinguish species can readily be recognised, which testifies to the considerable skills of the artists. In some cases, illustrations are stylised composites of several specimens.

One of the major failings of the 1844 work was not to provide adequate locality information for the specimens described, and so John Kelly, who worked for Griffith in the Valuation Office, published one such list in the *Journal of the Geological Society of Dublin* in 1855. In response, Griffith (1860) published a listing in the same journal and in the second edition of the Carboniferous *Synopsis*, published in London in 1862, he also provided locality data which took up sixty-two pages (Griffith, 1862). Both of Griffith's listings are far inferior to that of Kelly.

M'Coy completed his portion of the Silurian *Synopsis* in May 1845 and to this descriptive portion, Griffith had planned to append detailed geological accounts of the fossil localities. However, owing to his numerous public duties, the latter were not prepared. In addition, in the early part of 1846 Griffith collected, or had collected for him, more specimens from several localities, including Coolin, Co. Galway and Pomeroy, Co. Tyrone. From these, several new species were recognised by Salter, who had been given the task of selecting duplicate material for presentation to the University of Cambridge. This additional material was described by Salter and illustrated in a plate prepared by James De Carle Sowerby (1787–1871).

The Silurian monograph which was finally published in Dublin in the middle of 1846 contained seventy-two pages and five lithographic plates (numbers 1–4 drawn by M'Coy and number 5 by Sowerby). The taxonomic descriptions were similarly ordered to those in the earlier work: Cephalopoda, Gastropoda, bivalves, Brachiopoda, Crustacea (including trilobites), Entomostraca, Annelida, Echinodermata, and Zoophyta (bryozoans and corals). M'Coy erected the new fish species *Acanthalepis jamesii* based on a solitary scale, which is now regarded as a cystoid plate. The bulk of the material is decalcified or preserved as moulds and as such must have proved difficult to work with. All the descriptions of new species are short, while no information other than crude locality data is given for previously described species. Surprisingly, the later edition of 1862, issued in London, did not contain the detailed locality descriptions that Griffith had meant to prepare for the first edition.

No one knows how many copies of the two monographs were printed and how many were made available to the purchasing public. Of several copies seen by us, most were presentation copies distributed by Griffith to interested geologists or to institutions. While it is possible that few were actually sold, Griffith's tendency for self-promotion certainly resulted in the works receiving a wide circulation.

Conclusion

M'Coy has a reputation, probably with good foundation, as having been partial to 'splitting', or erecting numerous new species which today would be placed in one, perhaps morphologically variable, species. Nevertheless, a large number of M'Coy's taxa are still recognised as distinct, and so for students of Palaeozoic palaeontology, their descriptions and illustrations and the specimens upon which they are based remain important and salient sources of primary data. The two monographs of Irish Carboniferous and Lower Palaeozoic fossils produced by M'Coy in his youth in Dublin continue to be his greatest memorials and far outweigh those of his many medals and titles acquired later in life.

Note

A comprehensive assessment of Griffith's fossils and M'Coy's publications by Nigel Monaghan is in press: 'The fossil collections of Sir Richard John Griffith (1784–1878)'. *Irish Journal of Earth Sciences*, volume 43 (2025).

11

PATRICK GANLY (1809–1899) AND THE DISCOVERY OF EVIDENCE OF WAY-UP IN ROCKS ON THE DINGLE PENINSULA

The Kerry Magazine 7 (1996), 8–9.

Introduction

Since the late 1700s when geology as a science first developed, concepts and theories that are now readily accepted by modern geologists were formulated in various parts of the world. Geologists in Ireland and Britain made important contributions to our understanding of the sequence of rocks, their age, and the processes that formed them (Rudwick, 1985; Secord, 1986; Oldroyd, 1990). Several geologists who traversed Ireland's rocks are remembered today for their pioneering work. One man, Patrick Ganly (1809–1899) ranks high amongst them.

The Dingle Peninsula is one of the classic areas of Irish geology which has been extensively studied for nearly 200 years. Many researchers have written about its geological foundations which range in age from the 425-million-year-old Silurian fossiliferous beds and lava of the Dunquin Group exposed near Dunquin and Caherconree (Holland, 1988; Parkin, 1976), to the youngest Carboniferous limestones, 330 million years old, that fringe its northern margin between Tralee and the Magharees. Sandwiched between these two horizons are the sandstones and conglomerates of the Dingle Group and Old Red Sandstone which measure several kilometres thick and form the formidable mountain ranges of Slieve Mish and Brandon (Capewell, 1965). It was on the Peninsula that one of the fundamental concepts in structural geology was first proved—that of the use of cross-stratification in determining the correct way-up of strata, i.e., the original disposition of older layers of rock overlain by younger layers.

Patrick Ganly was born in Dublin and at an early age took up employment under Richard Griffith in the Boundary Survey in 1827 and later as a surveyor in the Valuation Survey Office until 1860 (Archer, 1980, 1985). Griffith's great ambition was to construct a geological map of Ireland and instructed his surveyors to collect geological data while engaged in their normal duties (Herries Davies, 1983). For eleven years between 1837 and 1847, Ganly spent the majority of his

 DOI: 10.4324/9781003475552-14

time employed on this unofficial geological survey. He travelled throughout the country and sent letters back to Griffith in Dublin that outlined the geological features and structures that he had seen. All this information was compiled and published in 1838 on a small-scale map by the Railway Commissioners of whom Griffith was one. The following year a larger map, at the scale of a quarter of an inch to one mile, was published and this went through various editions until the final version of 1855 (Herries Davies, 1977a). Griffith closely guarded the fact that he relied heavily on Ganly's observations in the construction of 'his' map (Herries Davies, 1983).

In the early years of the 1830s, Ganly was engaged in fieldwork in County Donegal where he examined sediments in a river bed two miles north of Carndonagh. These were formed into sand ripples, which could be seen on the sediment surface and in cross-section on the river bank where the layers of sand had been exposed by the downcutting action of the river. Ganly noticed that the cross-sectional shape of these ripples consisted of alternating troughs and crests. The ripples were asymmetrical in shape with a low-angled limb rising to the crest, beyond which was a steeply dipping limb (the lee face) on which sediment was deposited at a high angle (in what are now known as foresets). Ganly observed that this lee face was found on the down current side of the ripple. Such asymmetrical ripples are now known to be characteristic of a unidirectional flow of water, such as that produced by a river. He also sketched partially eroded ripples piled one on top of another, with the foresets accumulated in a stack of concave layers now known as cross-stratification.

Eight years later in June 1838, Ganly was at Coosnagloor in the shadow of Mount Eagle on the Dingle Peninsula, where he examined a sequence of steeply dipping sedimentary rocks (of the Dingle Group) exposed along 200 yards of the shoreline. These rocks appeared to be younging southwards throughout. At the northern end of the cliff exposure, Ganly found fossil ripples or cross-stratification similar to those he had seen in Co. Donegal, which indicated that the beds were indeed southward younging, but at the southern end, he found sediments where the cross-stratification was clearly upside down. This led him to conclude that the structure in the cliff was not a simple gradation from older to younger rocks but was a large-scale synclinal fold where limbs of older rocks were drawn together around a core of younger strata (Figure 11.1). Without the evidence provided by the cross-stratification, this conclusion would have been impossible to reach. He had determined that the orientation of cross-bedding could be used to determine if beds of sandstone were in their original position or had been turned upside down.

It was not until 1856 that Ganly published his findings. In June of that year, he addressed a meeting of the Geological Society of Dublin whose members received his findings with little enthusiasm (Archer, 1980) but nevertheless agreed to publish them in the Society's *Journal*. His short three-and-a-half-page paper which described a fundamental tool for unravelling the structure of folded rock successions is arguably one of the most important published in that periodical or indeed in any Irish geological serial publication. It was, unfortunately, ignored by the

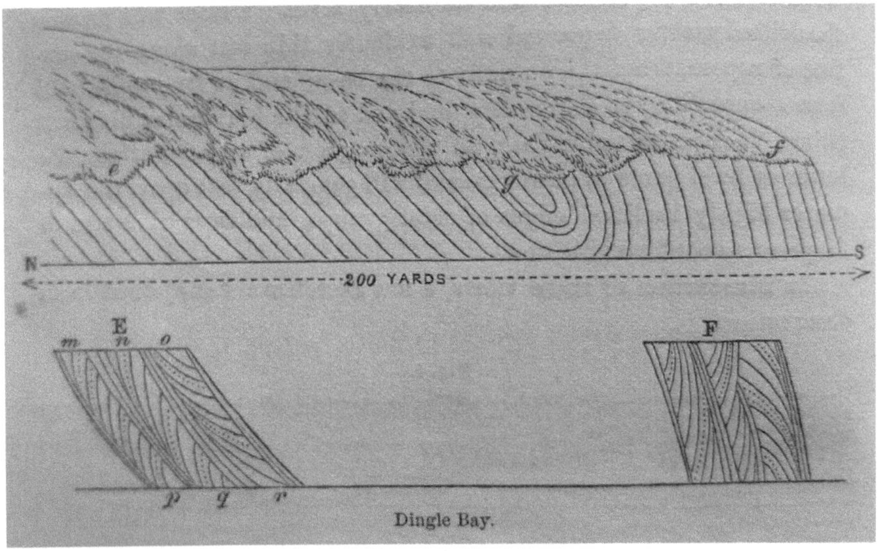

Figure 11.1 Original illustration from Ganly (1856, figure 2) of the cliff section at Coosnagloor showing highly inclined strata folded into a tight syncline. Beneath, the structure of individual beds is enlarged, and the disposition of cross-stratification shows that the beds on the right-hand side are younging to the left and are inverted (Geological Museum, Trinity College Dublin).

wider geological community and it wasn't until the second decade of this century that geologists working in the United States of America chanced upon the same method for demonstrating way-up in sedimentary rocks (Archer, 1980, 1985).

Ganly died in Dublin in 1899 and remained a little-known figure amongst historians of geology until the work of Simington and Farrington (1949) and Jean Archer, in particular, in the 1980s brought his contribution to structural geology to a wider audience.

Ganly's use of cross-stratification to determine the correct orientation of a sequence of rocks is today used by geologists throughout the world. Their use of it, pays tribute to a man who first used the method in a wind swept corner of the Dingle Peninsula.

12

A LETTER FROM WILLIAM BOOKEY BROWNRIGG TO THOMAS HENRY HUXLEY, DATED 29 NOVEMBER 1865, AUTHORISING HIM TO DESCRIBE HIS FOSSIL VERTEBRATES FROM JARROW COLLIERY, CO. KILKENNY AND GIVING DETAILS OF HIS FIND

Irish Journal of Earth Sciences 29(2011), 19–22 [co-authored with Migue deArce and Nigel T. Monaghan].

Introduction

The discovery of the Upper Carboniferous (Pennsylvanian) vertebrate fossils from Jarrow Colliery, Castlecomer, Co. Kilkenny in the 1860s remains one of the most important geological discoveries made in Ireland over the last 200 years.

The fossils were discovered by William Brownrigg of Brannockstown, Co. Kildare (Figure 12.1), who was a graduate of Trinity College Dublin and had a semi-professional interest in geology. He did not make a living from his geological expertise but he had been and continued to be an active member of the Geological Society of Dublin (and its successor the Royal Geological Society of Ireland) and was also a keen fossil collector. He had informally named one of his fossils *Wandesfordii*, after the owner of the land in which the colliery was situated, and had guessed that they could be reptiles, reckoning that he had discovered 'six or seven' new forms (Brownrigg, 1865).

The discovery of these fossils prompted rapid efforts to describe them, and priority over the finds, and hence the right to publish on them was contested by various scientists, both amateur and professional. Edward Perceval Wright of Trinity

DOI: 10.4324/9781003475552-15

Figure 12.1 William Bookey Brownrigg (Author's collection).

College Dublin, who had been approached by Brownrigg for assistance, turned himself to Thomas Henry Huxley, the undoubted contemporary expert on fossil amphibians, with a request to collaborate on their description. Correspondence between Wright and Huxley has been recently shown to be somewhat cautious, with both men anxious to gain the upper hand in recognition for the find and description of the fossils. These communications are discussed in some detail in DeArce *et al.* (2011).

In 2010, a manuscript letter from Brownrigg to Huxley, now contained in the Thomas Henry Huxley collection at Imperial College London [ICL11.125], came to light, and this adds to our understanding of the affair. Brownrigg's letter was in response to one from Huxley dated 29 November 1865, in which he wrote that he wanted to make sure that Brownrigg was happy that he would describe the fossils with a view to their publication. [Huxley's letter to Brownrigg is mentioned in archival item ICL29.139, a letter to Wright dated 29 November, and mentioned to Wright again on item ICL29.141. Huxley had read Brownrigg's reply on 30 November, and on the same date, he wrote to Wright with the news (ICL29.145).] As this until now unpublished letter provides important information on the background to the description of the vertebrates from Jarrow Colliery, Co. Kilkenny. What follows is a full transcript of Brownrigg's reply to Huxley's query, written from his townhouse in Dublin. The words in brackets are unclear in the original.

Transcript:

18 Adelaide Road Dublin
Nov. 29: 1865

My Dear Sir,

I hasten to assure you that you are most welcome to name, describe and illustrate all or any of my coal fossils which you said (could?) occur here. And I need hardly say that it gives me very great pleasure to think that those I collected and so saved from the fire are examined by you as worthy of so much trouble.

At the same time perhaps it is as well to give you a short account of my share in the transaction—About one year ago I found pieces of Wandesfordii at the Jarrow pit colliery with other fossil reptile remains feeling sure that I had never seen anything like them before. I promptly (went) up with them to Dublin and showed them from time to time to different friends but nothing satisfactory was made of them. About august last my friend Dr EP Wright was about describing some new insects which I gave him from the same coal measures and came to Castlecomer to look over my fossils found during the (scour?) for insect remains. He then told me he was so satisfied that I had a lot of undescribed reptilian and fish remains. that he would get a grant of money at the next—September meeting of the Brit. Assoc. to assist me in collecting more speci-mens. Eventually Dr EP (to) got me twenty pounds from the Brit. Assoc and came with me to Jarrow to explore the beds in which the fossils were found, and I then asked him to undertake the description of all my material and he consented on the conditions that nothing but collecting and drawing was to be (thereafter?) up till January and that he then should be at liberty to consult you on subject going over to show you the drawings to tell you the conclusions he might have come to, and ask you to join him in a memoir for the Academy or the Palaeontological society. Shortly afterwards Professor Haughton got some specimens which he said he was going to send at once to you to describe. Dr E.P.W then told me that it would be unfair to you to let you describe new forms on the few remains in Prof. H. hands, he (Dr W) having at the same moment the key to the whole in his fossils and from what he has since told me I believe he at once wrote about my specimens to you. I believe it to be his wish that you would describe them and I feel they could not be in better hands, at the same time fresh specimens will probably be coming to light (next) week and if you were both working together on the subject I thought Dr W prefers might be of some assistance to you he being on the spot and some such arrangement would I felt be pleasant to Mr Dobbs the agent of the mine who knows Dr W I know that Dr W is as quixotic as it is possible to be about taking the work from any one still more from one whom he thinks the best qualified to do it (he like you urged me to do it myself and he promised me to help and told me to be sure of your assistance too as he said you never failed to help on a young beginner, the truth of which from your kind offer I am now deeply sensible of) but I cannot think I do him any injustice in mentioning the (about) trusting you will excuse my going into these details and assuring you that your letter has completely removed from my mind any slightest feeling of irritation at my plans not being carried with exactly as I wanted, that may have existed. I want only again add that the fossils are for describing quite at your disposal. I remain my dear Sir yours very faithfully,

W Brownrigg.

PS I should have written yesterday but have been laid up since Sunday with a severe cold and attack of diarrhoea.

Discussion

Brownrigg acted in haste, writing and posting his letter immediately after reading Huxley's. Huxley acknowledged receiving it in a letter to Wright dated 30 November. Although Brownrigg's handwriting does not betray any obvious haste, some words are unreadable and his grammar and wording are obscure at times. Six times in the 583 words of text, Brownrigg gently asserts his ownership of the fossils and the fact that he found them and saw their novelty. E.P. Wright was welcomed to Castlecomer as a visitor after the existence of fossil vertebrates had been established. Clearly, in Brownrigg's mind, this ownership referred not just to the pieces of coal bearing the fossils, but to the intellectual act that their sight elicited, in his mind, of having identified them as new and valuable to science. Huxley's request was an acknowledgement of this priority. Only this interpretation, and the assumption that it must have been so obvious to all that no honourable person would consider twisting its meaning, can explain Brownrigg's consent to Wright's wish that Huxley would describe them. With his generous concession to Huxley, his dismissal of Samuel Haughton's plans and his acceptance of Wright's conditions, which amounted to a moratorium in reporting until after January, Brownrigg must have thought that he had created a climate of total trust between the three men; himself, Wright and Huxley, where goodwill would flow from all to all. In his letter to Brownrigg (ICL29.141), Huxley was explicit, asking for permission to describe and publish. Even if he was overawed—as he obviously was—at being approached by a world authority such as Huxley, Brownrigg specifically did not mention his permission to publish. This was immediately taken for granted by Wright and by Huxley. Huxley's rapid reaction to Wright's offer and his counter-offer of coming to Dublin immediately was not what Wright had expected. It was little short of an invasion, which Wright was unable to fend off. Wright had sent photographs of the fossils, and was prepared to send drawings, but eventually had to receive Huxley to inspect the fossils directly and then to send him the entire collection at Jermyn Street, London. Wright suggests that the paper to be published in the *Proceedings of the Royal Irish Academy* was still pending Huxley's approval of Wright's conclusions, but he was already committed to it before an unsuspecting Brownrigg. While Brownrigg barely disguises his unease he does agree to Huxley's request, no doubt realising that the fossils deserved the highest expertise available for their description.

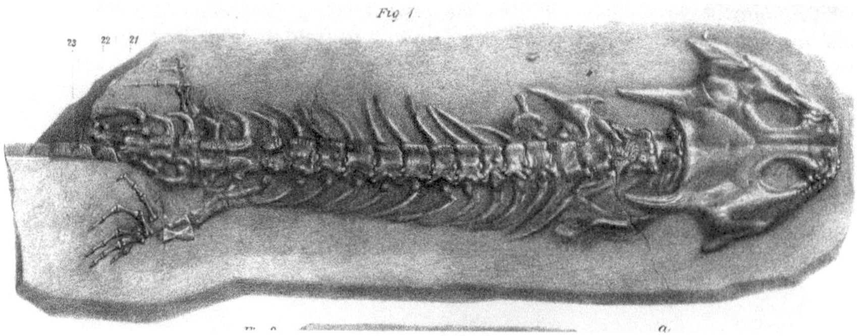

Figure 12.2 *Keraterpeton galvani* Huxley, an amphibian from the Pennsylvanian Coal Measures of Jarrow, Co. Kilkenny, Ireland. Lithography by Joseph Dinkel in Huxley and Wright, 1863 (Author's collection).

The following year Wright and Huxley (1866) published a short paper that heralded an authoritative paper of five years later, in which Huxley solely provided the full taxonomic descriptions of the fossils (Huxley and Wright, 1871) (Figure 12.2). This monographic paper was published in the *Transactions of the Royal Irish Academy* to which Wright had contributed the section on the geological setting of the find. Much of Wright's information had come directly from the observations of Brownrigg who had earlier published a brief outline of the discovery (Brownrigg, 1865).

Following the publication of the 1866 paper by Huxley and Wright, William Hellier Baily of the Geological Survey of Ireland claimed, with questionable justification, priority over the finds (Baily, 1866). It is clear that Wright, an academic, sidelined Brownrigg in the whole affair even though as finder of the specimens he had expected to be involved in the scientific description of his finds. It is not clear what Brownrigg's immediate reaction was to the publication was, or whether he was appeased somewhat in having one of the fossils, *Ophiderpeton brownriggii* named for him by Huxley. It was not unknown for collectors of important material to be marginalised by academics and others when it came to publishing reports on material (Torrens, 1995; Knell, 2000). Shortly afterwards, on account of due reflection of the episode that must have left him feeling disaffected and aggrieved, Brownrigg abandoned his study of geology altogether (DeArce *et al.*, 2011) and he retreated to his small estate in Co. Kildare.

The Jarrow Amphibians constitute one of the most important vertebrate faunas from the Pennsylvanian. Aside from Huxley's original treatise, they have been studied continuously up to the present day (see, for example, Milner, 1980; Sequeira, 1996; Milner, 2019; Ó Gogáin and Wyse Jackson 2021). Brownrigg's fossils are now housed in Trinity College Dublin and the National Museum of Ireland (Wyse Jackson and Monaghan, 1995) and further Jarrow specimens are in the collections of the Geological Survey Ireland, the Natural History Museum London, and Manchester Museum.

13

SERVANTS OF THE RAJ AT PLAY: AN EXAMPLE OF THE SOCIAL INTERACTIONS BETWEEN A GEOLOGIST AND AN ENGINEER IN NINETEENTH-CENTURY INDIA

The Compass 81(1–4) (2008), 96–103.

Introduction

Biographical studies can provide a powerful insight into the social and scientific interactions between members of scientific and professional communities. However, as time passes from the death of an individual to the time of study, it becomes increasingly difficult to acquire information. In many cases, the only information available to the researcher is official communications and publications from the individual and often rather sanitised obituaries. Manuscript sources are more difficult to find or may simply not have existed. While a picture can be assembled of an individual's career and activities what is rarely revealed are the day-to-day happenings of that individual, their character, their likes and dislikes, the way that they relaxed, and sometimes intimate family details are non-existent.

Through the use of available manuscripts and personal sources, it may be possible to build up a fuller character assessment of the individual or group of individuals under study. At the present time, we tend not to keep correspondence, or else we correspond electronically. Some scientists and engineers or members of their immediate family in the past kept personal diaries, journals, or daybooks and these may provide invaluable information that when combined with more readily available information can allow for a greater visualisation of the social circles in which the persons lived, and a fuller understanding of the various linkages between them and their contemporaries. Such a study has highlighted the problems encountered by Charles Lyell and Roderick and Charlotte Murchison while undertaking fieldwork in France and Italy in 1828 (Kölbl-Ebert, 2007). If possible such sources of information should be sought and if found utilised by historians of geology, so that the history of the subject and the contributions of early geologists can be placed in a fuller context.

 DOI: 10.4324/9781003475552-16

In the 1990s and early 2000s, a number of studies have examined the development of the Geological Survey of India (Sangwan, 1994; Kumar, 1995; Grout, 1996; Leviton and Aldrich, 2000). These are valuable academic studies but tended not to relate the day-to-day activities of the geologists and difficulties that they had to overcome.

Charles Allen (1975) in his *Plain Tales from the Raj* has provided using oral and other personal histories, fascinating information on India under British rule between 1900 and independence in 1947, and what it was like to live in the subcontinent for members of the civil service.

However, what is required are investigations of the effects of colonial rule on the indigenous population and how geological and engineering activities in the 1800s affected the population and still impact Indian life.

In this chapter, examination of two professionals in India throws light on the close-knit nature of the scientific community and their day-to-day activities, both professional and recreational. It is clear that the participants lived privileged lives.

The Geological Survey of India and its connections with Ireland

Irish geologists have frequently travelled abroad to work—a pattern typical of many Irish professionals. In the 1800s, Thomas Condon worked in north-western America, and Patrick Joseph Pentland worked in Peru. Andrew Wiley served as Government Geologist to the Cape Colony between 1855 and 1859; Gerrard Kinahan, son of an Irish geologist, was killed by a poisoned arrow while engaged in fieldwork in Niger in 1886; and Edward Townley Hardman was involved in the discovery of gold in Western Australia.

However, by far, the largest migration from Ireland for official government service was to the subcontinent of India. Trinity College Dublin saw an opportunity and filled an educational niche and in the 1850s under the geologist Samuel Haughton and economist Joseph Allen Galbraith developed a school for the training of professionals for the service of the Indian Civil Service. This programme produced geologists, surveyors, naturalists, and engineers, and it is the interrelationships and friendship of two such graduates that this chapter focuses on.

In 1851, the Geological Survey of India was established (Chaudhuri, 2001) and Thomas Oldham, the Director of the Geological Survey of Ireland and Professor of Geology at Trinity College, Dublin, left Ireland with his new wife to take up an appointment as Superintendent of this new Survey. He subsequently employed a number of Irish geologists for service in India, offering them higher salaries than were available to them in Ireland. His brother Charles joined him in 1856 and his own son Richard joined in 1879.

At least nineteen men who served the Geological Survey of India were either Irish by birth, descent, or who were educated in Ireland. These will be the subjects of a future communication. Oldham, Henry Benedict Medlicott (who coined the term 'Gondwana Series') (Wyse Jackson, 2002b), William King (Anon., 1972),

Richard Oldham (who demonstrated the presence of the Earth's core) (Wyse Jackson, 2002c), and Henry Hayden were all Directors of the Geological Survey of India, while closer to the present day Henry Crookshank (1893–1972) was Director of the Geological Survey of Pakistan. For nearly seventy years, there continued a tradition of Irish involvement in the unravelling and understanding of its geological history.

The geologist and the engineer

Charles Æmilius Oldham (1831–1869) and Thomas Hardinge Going (1827–1875), both graduates of Trinity College, Dublin arrived in India from very different routes, but once in the subcontinent became firm friends. Charles, the youngest brother of Thomas, was a graduate of 1851 in ethics and logic and he later trained as a geologist at the School of Mines, Jermyn Street, London. He was appointed to the Indian Survey in April 1856 and rose to the rank of Deputy-Superintendent by 1862. He also held an appointment as a lecturer in geology at the Engineering College in Madras.

For a time, he was the officer in charge in the Madras Presidency (Fermor, 1976) where he worked alongside two colleagues Bruce Foote and his fellow Irishman William King, the son of William King, Professor of Geology at the Queen's College, Galway in Ireland (Harper 1988a, b). The Madras Presidency is in the south portion of India. Between 1857 and 1867 Oldham, Foote and King mapped 58,000 square miles (Anon., 1972) which was quite remarkable given that this is 1.5 times the size of Ireland which took 45 years to map.

Fieldwork was often difficult with the geological parties having to contend with working in isolated regions, difficult terrain, and they also were not immune from disease. Cholera, typhoid, and even sunstroke were not uncommon, with Foote so affected in 1863, and Oldham suffering the loss of some of his servants to cholera in 1864. Four geologists were less fortunate than Foote and they died in service between 1858 and 1861 all within a short time of their arrival in India (Fermor, 1976, p. 53).

Geologists generally conducted fieldwork during November to May (Ball, 1880) and were largely confined to headquarters in Calcutta or Madras during the remainder of the year as was either too hot or monsoonal for work outdoors. As a consequence, they spent long periods away from home. While on fieldwork they stayed either in bunglows (single story dwellings) or in tented accommodation. Ball (1880, pp. 8–9) gives us a flavour of what a fieldcamp was like:

> I found my colleague encamped in a picturesque grove near the road, and when my own tent was pitched I had an opportunity of forming some idea of what my future life would be like. And here it will be convenient to enumerate and describe the various items which are found to be requisites in the camp establishment of an Indian geologist The tent generally used is what is called a single-poled, double- roofed hill-tent,

> which has an internal superficial area of twelve feet square; and two lateral verandahs which can, if desirable, be enclosed by walls—thus making two side rooms. Small as this area may seem, it is, with a little management, capable of containing all requisite furniture and storage space. The furniture generally consists of a frame bed (or charpoy) with plaited tape; two tables, one for eating purposes, the other for writing materials and for supporting a travelling book-case; these, with a couple of chairs, gun rack, clothes hooks, and sundry portmanteaus and boxes, make up the sum of the internal equipment Many people who use tents of this size carry a second for their office, and for the purpose of sending on ahead to the new stage; while district officers, except the most junior, generally use several larger and much more commodious tents. However, for a geologist, most of whose time is spent out of doors, and who has no regular office work involving formal intercourse with crowds of natives, such a tent is amply sufficient for all purposes, and is certainly the best in the hilly regions where carriage is difficult. This form of tent, moreover, if properly made of good materials, is wind and weather proof, and will withstand storms which would lay low most others.
>
> A bath tent, and three sleeping *pâls* for the native Doctor, the kitchen, and the Jemidar and Chuprasies [two grades of servant] respectively, are further requisite; though some people consider the shelter of a tree sufficient for a kitchen, and the huts which the villagers run up with branches as ample protection for the servants. Most Indian servants, if provided with a blanket, think it no great hardship to sleep in the open; but I am fully satisfied that the tale of sickness is much diminished by the provision of proper covering as a protection from the cold by night and the sun by day.
>
> As the geologist requires no clerks or writers, and as until recently there were no native assistants, the list of his followers resolves itself into the following elements:—One native doctor or trained hospital assistant, who is in charge of the health of the men, and is generally competent to deal with the ordinary forms of disease and simple surgical cases. One Jemidar or headman, under whose orders all the rest of the men are supposed to be; his pay is from 10 to 16 rupees per month, and his perquisites, as he keeps accounts of the general expenditure, are no small source of additional in come to him. Six Chuprasies, who carry a *chupras* or engraved brass plate on a sash or belt, indicating the service to which they belong. The duties of these men are multifarious: they act as tent pitchers, messengers to post, orderlies and night watchers; they also obtain the supplies from the villages, going on a day in advance in order to make the necessary preparations. The pay of these men is from 6 to 8 rupees per month each.

Charles Oldham and his colleagues mapped an Archean basement, overlain with thick successions of Cambrian sandstones, limestones, and shales, with younger

Mesozoic and recent sediments found along the coastal strip. The Mesozoic sediments included laterite and Jurassic plant-bearing successions. Chromite, and magnetic iron-bearing rocks were discovered and the indigenous methods of extracting the latter metal were described in a Geological Survey Memoir by Foote. Oldham discovered Jurassic plants at Nellore which significantly delineated the southernmost extent of the Gondwana Series in India (Fermor, 1976). However, perhaps the discovery that generated most excitement in the scientific community was that of stone implements in the area to the north and west of Madras which were contained embedded in lateritic conglomerates. These were found in 1863 and 1864 by Foote and King (Foote, 1868) and similar finds were found elsewhere by other Survey officers.

Thomas Hardinge Going (Figure 13.1A) was born in 1827 in Cahir, Co. Tipperary, Ireland and graduated the year before Oldham. He was to find work as a railway engineer first in Pennsylvania and Arkansas, before taking up a position with the Madras Railway Company in 1857. He arrived in India on 11th May 1857 on the S.S. *Ava* and embarked on a survey of the proposed north-western railway line from Madras [modern name: Chennai] to Raichore [Raichur]. He also surveyed a possible route to Hyderabad in 1862 but this line was never built. Then, during an eighteen-year period (1862–1875), he constructed the Madras to Raichore railway (Going, 1984). He had to overcome major difficulties caused by the difficult terrain in which he had to work and problems with the wooden sleepers that had to be replaced with cast iron sleepers. Many lives were lost to cholera and other diseases in building this railway, and occasionally, the physical structure of the railway itself failed with tragic consequences. In 1870, the bridge over the River Cheyair collapsed and several lives were lost (*Madras Mail* 24 October 1870). In 1871, Going's Madras to Raichore line joined up with that being built south-eastwards from Bombay [Mumbai], and travellers were able to cross India from east to west for the first time. Going's engineering achievements were of great significance.

Social lives and recreational activities in India

Thomas Going kept a diary for many years, and these are now in the British Library. These provide a rare insight into the social and scientific interactions between some members of professional community and their families in India at this time. They demonstrate that Going developed an interest in natural history and that he was encouraged by his friends to collect minerals and fossils and zoological specimens. He donated material to the Madras Museum and a number of zeolite minerals collected and described by Samuel Haughton (1871b) by him are now in the Geological Museum in Trinity College, Dublin (Figure 13.1B).

In the period around 1861 and 1862 when Going had most contact with the geologists from the Madras Presidency, he was working in and around Pootore near the Tripetty Hills. Tripetty is in an area 170 km north-west of Madras and contains many impressive temples which today attract a large number of pilgrims.

Figures 13.1 (A) Thomas Hardinge Going; (B) Stilbite collected by Going from Bore Ghaut, India, now in the Geological Museum, Trinity College Dublin.

When they had time for recreation, they picnicked, went sketching, hawking, riding, and shooting. They played shuttlecock (badminton), went boating and walking in the hills. When in Madras, Going and Charles Oldham would meet for dinner at their Club. The diaries illustrate the close-knit nature of this small and fragile community, which was bound closer still when in late 1863 Oldham married Evelyn King, the second daughter of Professor William King and sister of his geological colleague (Anon., 1869). Going was unable to attend the wedding. The couple had one son, but Going's diaries also suggest that they had another child who died in mid-1867, after which Oldham moved into the Madras Club. It is probable that Mrs Oldham returned to Ireland for some time after the child's death.

Below are some extracted entries from Thomas Going's diaries which illustrate the close bonds within this group, document some of their leisure activities, and illustrate the scourge of illness that affected Europeans working and living in India. My annotations are in square brackets.

1860

Sept 1–7: Foote has come up on sick leave and his Colleague King is also labouring under illness. He gives me news of the Meddlicots [sic Joseph and Henry Medlicott who both worked as geologists with the Survey] and also of Oldham and I find from him that an order has been issued for the examination of the mineral-bearing rocks of Cuddapop District so that we may expect a visit from the geologists. Foote and I employed ourselves looking over my collection of Zeolites and dining together at the Club.

Dec 14, Thursday: Breakfast with Foote and King and was introduced to his sisters [one of whom, Evelyn, later married Charles Oldham]. Then called at Club and saw Oldham who is coming up my way in a few days, with King. We have arranged that the contemplated excursion to Naggery Nose [local sandstone hill topped by a fort] will be commenced on the day after Christmas. The Ladies &c will come up to Pootore on that day to dine, by evening train. [Pootore is a small town close to where Going was working and living at the time.]

Dec 22, Sunday: Oldham and King came to breakfast. They returned to Pootore after dinner to be ready for the morning train to Madras, to attend Ball of 43rd [Regiment].

Dec 24, Tuesday: Oldham, King & Pinson [an engineer] arrived this morning.

Dec 25, Wednesday: Had a Camp pitched.

Dec 26, Thursday: Mr Percieval [an Anglican minister] and his two daughters came up by evening train with Foote and his sister and Miss King [it is unclear if this was Evelyn or the other sister]. We all went out to the fort at Nag. Nose and had dinner at 11 o'clock [at night].

Dec 27, Friday: Started at 7 this morning and reached top of Nose. Some at $10^1/_2$ and the rest at 11 o'clock. After breakfast, we had some sketching, singing, chatting and exploring. Started to descent at 3 pm and reached Camp at 6 to dinner. A successful day.

Dec 28, Saturday: All rode across country to the Railway and arrived at Porly at $10^1/_2$ to breakfast. Ladies took possession of the bungalow—Gentlemen of the tents. All highly pleased with the excursion and the country. Strolled around in the evening. Very pleasant dinner party in the tents.

Dec 29, Sunday: Mr Percieval performed service in the verandah. Ascended the hill behind the Bungalow an opportunity which Foote availed himself of to propose to Miss Percieval who has accepted him and pleased all parties. [She was to die on 30th June 1870 of abscess in her liver.]

Dec 31, Tuesday: Had a long ride this morning with Miss King and Robinson. Pistol practise in the evening and shooting teal. We all sat up and saw in the New Year very pleasantly.

1862

Jan 1, Wednesday: Went out with Oldham this morning and bagged some teal. Numerous presentations of Fruit and flowers today.

Jan 3, Friday: Oldham and I killed 2 Cobras near the Bungalow. One measured 6 ft 9 in. a monster. Miss K & I rode with Robinson.

Jan 4, Saturday: Miss King unwell.

Jan 6, Monday: She [Miss King], Oldham and self-rode.

Jan 14, Tuesday: King and his sister rode out with me to breakfast at Poody. Jan 15, Wednesday: Thermometer up to 85° today,

Jan 17, Friday: My horse is lame in consequence of being ridden by Miss K without a shoe.

Feb 5, Wednesday: King and his sister moved to Camp at Yarepet. Very lonely after parting with my friends.

Feb 21, Friday: Played Battledon and Shuttlecock with Miss King.

March 21, Friday: Called on the Percievals [at Madras] with Roberts. Miss P told me that Doc. [Thomas] Oldham [Director of the Geological Survey of India] refused to allow Foote to be married before he finishes the sheet he is working on. In the morning called on Museum and left my bottle of spiders &c. Capt. Mitchel was not there.

March 25, Tuesday: Letter from King who says Foote's wedding is put off to early in June.

April 27, Sunday: Oldham arrived to Breakfast. King and his sister arrived to dinner.

May 1, Thursday: We all started at $4^1/_2$ this morning and reached Pootore for the 6.20 train.

June 7, Sat: Foote and Miss Percival married at St. Thome [Thomas'?] Church.

June 9, Monday: Stevenson, an officer who was at the wedding on Saturday has died of cholera.

June 18, Wednesday: King arrived to breakfast. He remains for 2 or 3 days and then goes to Madras to bring his sister to the Hills.

Oct 22, Wednesday: Oldham tiffed with me ['tiffin' is an Indian word for luncheon]. Nov 14, Friday: Had a game of croquet with Oldham.

Dec 7, Sunday: Oldham rode out this morning & we had a walk in the Chalk Hills.

1863–1868

Minor entries relating to Oldham and family.

A natural but early end to the friendship

Working in India was not easy. As we have read a number of geologists died as a direct result of their labours in India while others suffered from poor health: Thomas Oldham retired on such grounds in 1876 and died only two years later, while his son Richard suffered from the tropical disease sprue (Wyse Jackson, 2002b).

Oldham had returned on furlough with his family to Ireland in December 1868, and by early 1869 was showing signs of serious illness. On 30 March 1869 after a very painful illness, he died at his brother's house on Wellington Road, Dublin, at the age of 38. He had succumbed to blood poisoning caused by the decay of a guinea-worm that had burrowed into him in India. He left a widow and a young son; she died many years later on 18 May 1919 of heart failure, on board the S.S. *City of Karachi* in the Red Sea either on her way to or from India.

Thomas Going became gravely ill in early 1875 in returned to Ireland with Louisa, née Hely, his wife of two years and two young sons. He died at the age of 48 at Killiney, south of Dublin on 15 September of that year. By a curious twist of fate, both men died not in India but in Ireland and are buried close by each other in Dublin's Mount Jerome Cemetery. Both died young as a direct result of illness contracted in India.

Conclusions

Life in India in the service of the British Empire in the nineteenth century was complex, particularly for those men who worked largely outdoors. They were subjected to working in remote areas where they were open to disease and often death. Two such men working in the Madras district as a railway engineer and as a geologist made considerable contributions in their fields. They became great friends, as testified in the unpublished personal diaries of Thomas Going, the engineer. These reveal the closeness of the professional community in Madras, which entertained themselves through dining at their Club, picnicking, playing games or riding.

Studies of the day-to-day lives of scientists through examination of diaries and journals and other non-published, non-official materials, can provide a fuller understanding of the conditions under which they worked and can allow historians of geology develop a greater understanding of the social interactions of their subjects, and the influences that came to bear on their professional work and activities.

However, all these insights need to be placed alongside the impact their lives had on both the Indian population and the Indian economy that was developed and exploited to enhance the lives of the ruling classes and the wealth of the colonial power.

14

WILLIAM HELLIER BAILY (1819–1888): FOREVER AN ACTING PALAEONTOLOGIST WITH THE GEOLOGICAL SURVEY OF IRELAND

The Geological Curator 9(2) (2009), 57–84 [co-authored with Matthew A. Parkes].

Introduction

By the mid-1840s, it had become clear to men of science that systematic study of fossils was of fundamental importance in deciphering the geological message revealed in successions. From the outset of the Geological Survey of Ireland's (GSI) mapping of the country in 1845, the importance of collecting fossils to correlate the strata and assign ages to the various formations was recognised as critical, and staff to support such endeavours were appointed. The staff of the fledgling organisation included two fossil collectors and also Frederick M'Coy who was appointed an Assistant Geologist. He had considerable palaeontological experience despite being only twenty-two years old, having catalogued the collections of the Geological Society of Dublin (M'Coy, 1841) and been in the employ of Richard Griffith (1784–1878), who had produced the first large-scale geological map of Ireland in 1839 (Herries Davies 1983, 1995), M'Coy described the fossil collection now known as the Griffith Collection in two monographs on the Lower Palaeozoic and Carboniferous fossils of Ireland (M'Coy 1844, 1846; Wyse Jackson and Monaghan, 1994 [Chapter 10 herein]). M'Coy's time in the GSI was not a success and he resigned after only sixteen months (Herries Davies, 1995, p. 315). He later became Professor of Geology at Queen's College, Belfast before moving to Melbourne in Australia where he had a glittering career. The senior Fossil Collector was James Flanagan, who had worked for the Ordnance Survey of Ireland and with Joseph Ellison Portlock in Londonderry, where he helped assemble an important suite of materials (Portlock, 1843; Tunnicliff, 1980).

The GSI was a branch of the Geological Survey of Great Britain (GSGB) or Geological Survey of the United Kingdom, as it was named at different times,

DOI: 10.4324/9781003475552-17

and the Palaeontologist provided determinations of the specimens collected by the field geologists and advised them on the stratigraphical implications of the finds. First Edwards Forbes (1815–1854), then his successor John William Salter (1820–1869) paid visits to Dublin to examine the collections made during early mapping years.

However, despite the renowned, and even illustrious, palaeontological names associated with the GSI's fossil collection, that of William Hellier Baily (Figure 14.1) carries an enduring quality and a hallmark of distinction without his having a single volume of the quality of M'Coy's synopses, Portlock's memoir of the geology of Londonderry and adjacent areas (1843) or their English counterparts.

Baily, who served as the Acting Palaeontologist from 1857 until his death in 1888, is the subject of this study. Here, we explore facets of his background, character, family life, his scientific work, and in particular his contribution to Irish palaeontology through his publications and curation of the fossil collections in the Geological Survey. We also examine his relationships with some of his colleagues and assess the professional disappointments that he felt he had to endure throughout his time in Ireland.

Figure 14.1 William Hellier Baily (Wellcome Images. CC-BY-4.0).

Baily's ancestry and early life

William Hellier Baily's immediate ancestors are quite difficult to unravel because of the use of very similar names and variants. He is descended from Dr Walter Baily (1529–1592) a physician to Queen Elizabeth I (Ossie Baily, written communication to PNWJ 21 November 1996). Later, Bailys hailed from Frome Selwood (now just known as Frome) in Somerset, and some later became active in the commercial life of the seaport of Bristol situated close by.

His great-grandfather was a James Baily who died before 1811, and grandfather was William Hellier Baily (born 1763 Bradford-on-Avon, Wiltshire which is just north of Frome Selwood across the county boundary) who moved to Bristol and became a carver of figureheads for ships [He should not be confused with the man of the same name who was a tea merchant and grocer of 44 Redcliff Street, Bristol (*Pigot's Directory of Gloucestershire* for 1830)]. In later obituary notices of his son the sculptor, it was noted that William was one of the foremost exponents of his craft in the country. For a while, he was in partnership with a Mr Wolles and the pair occupied a premises on the Quay in Bristol until the partnership was dissolved in 1806 (*London Gazette* 14 July 1813, p. 1390). He married Martha Hodges in about 1787 and set up home at 1 Red Lodge Court. She was related in some way to Edward Hodges who in the early 1800s constructed some of the Martello Towers around the shores of Britain as a response to the Napoleonic threat. The couple had four children, two sons and two daughters. Their two daughters were Mary Ann (born 30 May 1790) and Martha (born 19 December 1793). Edward Hodges Baily (1788–1867) was their eldest son and he became the premier sculptor of his day in England. He was responsible for the statues of Nelson in Trafalgar Square and Bishop John Jebb in St Mary's Cathedral, Limerick in Ireland and, coincidently, also carved a marble bust of Henry De la Beche (see Sharpe and McCartney, 1998, pp. 149 and 252) who later gave his artistic nephew William employment with the Geological Survey. His daughter married his pupil Edgar George Papworth (1809–1866) who became a renowned sculptor in his own right.

William and Martha's second son, John (5 August 1792–1731 July 1868) followed his father into the trade as a carver and guilder of figureheads, and in 1830, he was living at 2 Park Place, St Michael's Hill, Bristol. This district of the city, it would appear from *Pigot's Directory of Gloucestershire* for 1830, to have been a haven for artisans and craftsmen. Later in life, John Baily became a commercial traveller (*The Bar Reports* 1869, pp. 278–279). He had at least three children: William Hellier Baily (born in Bristol 7 July 1819, possibly at 2 Park Place), the subject of this paper, and two daughters, Priscilla (born about 1825, died Bath, Avon 1875, aged 50) and Mary Jane (married 1874, Bristol). This is known because on 17 April 1869, a problem with his appointment of executors came before the courts (*The Bar Reports* 1869).

Baily's early forays into scientific work

After his schooling William went to work for a solicitor in Bristol but having developed his artistic skills from an early age, a long-term career in the law was unlikely. In 1837, when aged 18, Baily was taken on as Samuel Stutchbury's (1798–1859) Assistant in the Bristol Institution (Crane, 1983) and his scientific career began. Stutchbury may have utilised his young assistant's abilities as a draftsman, although as yet we cannot find any published lithographs or drawings from this period.

After leaving the employment of the Bristol Institution, Baily quickly found another opening with John Phillips, nephew of William Smith. The De la Beche correspondence in the National Museum of Wales includes letters from Edward Forbes and John Phillips from the mid-1840s which are illuminating. Baily, whilst clearly a junior fellow and subject to direction in his work, was also apparently liked as an individual and trusted to complete work satisfactorily, mainly drawing specimens for publications. Phillips, on 23 December 1843, sought De la Beche's opinion on the salary of £2/2/0 a week he proposed for 3 months work he was finding Baily starting January 1844. Later, on the 24 August 1844, he finished a letter writing:

> In regard to Baily, I have the strongest desire to help him, and if any difficulty should arise in respect of his employment, I should hope to be informed, as some thing may be in my power. If he were not employed on the Survey, I could give him a month's work or more immediately, on my own account, and would gladly do so.
>
> (National Museum of Wales (NMW) De la Beche archive 1562)

Later in April 1844, Baily was appointed to the GSGB as a draughtsman and promoted to a museum-based Assistant Geologist in April 1845 on a salary of 9 shillings and seven pence per day (= £149/19/7 per annum) which allowed him avoid paying income tax (British Geological Survey Archives GSM 1/7, p. 94). This took him to London where he was to purchase a house, at 18a Walcot Place East, Kennington Road in which he was living in late 1855 just before he moved to Ireland (*London Illustrated News* 1856). In 1854, he was appointed an Assistant Naturalist.

Whatever favours or patronage may have been instrumental in opening doors for Baily into the GSGB, his talents as a draftsman and former Museum Assistant in Bristol were clearly sufficient to establish himself, since later letters from A.C. Ramsay to De la Beche indicate that Baily's salary and income tax was being paid by De la Beche himself in 1850 (e.g. NMW De la Beche archive 1797).

In the Museum of Practical Geology in Jermyn Street that opened in 1851, he continued to produce drawings and engravings on stone for lithographs but he also came under the management of Edward Forbes until he moved to Edinburgh in 1854, and later under Thomas Huxley. No doubt much of his work would have been mundane identification and curatorial work, but this was valuable experience for a future palaeontologist in his own right, and this palaeontological experience was to later be put to great use in Ireland. He would have helped make material available to others for study and in 1848 provided some barnacles for Charles

Darwin to study (Darwin to Baily, letter 5 October 1848; letter number 1201, www.darwinproject.ac.uk).

The following year he produced a crest for the 'Royal Hammers' as the Geological Survey had informally become known. This took the form of a trilobite supported on the left-hand side by a dapper spectacled Mr Ichthyosaur and on the right-hand side by Miss Plesiosaur (McCartney, 1977, p. 59; Fara, 2005; see also Wilson, 1985 front cover for a redrafted version). From 1849, the geologists composed verses (or 'songs') which were recited at the Annual Dinners; Baily contributed four such songs: 'Lament of the Irish elk' (1856), 'Ode to the Megatherium' (1856), and 'The Geological Surveyer' (1857) (GSM 1/146).

By the mid-1850s, the young scientist had gained enough experience to be entrusted with some research projects of his own and by 1859 he had published two papers, both in the *Quarterly Journal of the Geological Society* of which he was to become a Fellow. [A full bibliography of Baily's published papers, his contributions to the British Association for the Advancement of Science (BAAS) and the GSI *Memoirs*, and his artistic work for the Geological Survey and for the publications of other scientists is given in the original version of this paper, pages 80–84.] The first of Baily's publications described some Cretaceous fossils from South Africa (Baily, 1855), and the second was on fossils from the Crimea (Baily, 1858). The latter were also the subject of a letter by him published in the *Illustrated London News* in 1856. The Crimean War was waged between 1853 and 1856 and the material about which Baily published had been brought back to England by General Charles Frederick Cockburn (1830–1908). It is now in the Natural History Museum, London (Cleevely, 1983, p. 46). He was to return to the subject of fossils from the Crimea in a later paper that described the collections in Dublin (Baily, 1858–1859).

Baily as an artist and draftsman

Employed initially as a draftsman, Baily together with his colleague Charles Richard Bone produced drawings and the engravings on stone for plates to illustrate the figures and descriptions illustrative of the British Organic Remains series which was part of the *Memoirs of the Geological Survey of the United Kingdom* that appeared between 1849 and 1853. Baily and Bone were skilled draftsmen and undoubtedly their work took a considerable portion of their time.

Apart from his official Survey tasks, Baily was in some demand as an artist and he undertook some additional palaeontological commissions while in London. His work appeared in the some important geological and natural history journals and monographs in the early 1850s. Thomas Wright, a Cheltenham surgeon and natural historian, commissioned him to illustrate fossil echinoids from Britain and Malta that he published on in the *Annals and Magazine of Natural History* in 1855 (Wright, 1855; see also Zammit Maempel, 2007, p. 40). Similarly, he undertook work for Edward Forbes who had him provide six plates of bivalves and echinoids to partially illustrate his major paper on the fossil invertebrates of southern India published in the *Transactions of the Geological Society* (Forbes, 1845), and later more plates for his monograph on the naked eye medusae published by the Ray Society (Forbes, 1848). Both Forbes and Wright as well as John Morris had him illustrate their monographs published by the Palaeontographical Association.

Once he arrived in Ireland, he continued his artistic work in an official capacity for the Geological Survey that involved often engraving on wood the illustrations for the Memoirs that described the geology illustrated by the one-inch to one-mile geological maps (see Parkes *et al.*, 2000). He continued to work for others including William King of Galway (1857), a Mr Waller (1860), the Rev. Samuel Haughton (1860), Andrew Leith Adams (1863), and his GSI colleagues, Joseph Beete Jukes (1862, 1863) and George Henry Kinahan (1878). For Adams, Baily provided illustrations of two new species of the Maltese Pleistocene dormouse *Myoxus*. However, later the author complained that Baily's representations were not accurate and so he had new illustrations prepared by another artist and published in 1870 (Zammit Maempel 2007, pp. 48–49).

Can we sufficiently appraise the artistic talent of Baily? The work of George Victor Du Noyer (1817–1869), a contemporary Senior Geologist on the Survey, who started with Portlock's Ordnance Survey in Londonderry (Coffey, 1993), was appraised by Archer (1993) and brought to general attention in 1995 with an exhibition of his works in the National Gallery of Ireland which was accompanied with an excellent illustrated catalogue (Archer, 1995; Croke, 1995). His work has received more attention (Hegarty, 2017; Monaghan, 2001, 2017; Murray, 2017). The question could be posed as to why Baily has not received as much recognition. Clearly, he could paint as well as draw, shown by an uncatalogued large (*c.* 70 by 45 cm) watercolour painting in the GSI archives, but he possibly laboured at it rather than with the flair and speed of Du Noyer. Although he produced large drawings and paintings for Jukes for his College lecture series as did Du Noyer (Parkes, 2021), none are known to have survived in the GSI archives, and this is possibly the main reason he has not hitherto gained the reputation accorded to Du Noyer. There is absolutely no doubt that his talent lay in the precision and detail of accurate portrayal of the fossils, and in the detail of his lithographs and fine woodcuts.

The appointment of Baily to the GSI

After Salter's refusal to deal with his Irish responsibilities, Jukes was desperate in his need of a palaeontologist to progress the Survey and in a letter of 28 October 1856 and, in the 1856 Annual Report, made detailed pleas for assistance. The Director General of the Survey, Roderick Impey Murchison wrote to Earl Granville on 28 May 1857, arguing that as the BAAS meeting was due to take place in Dublin the following year that it was imperative that someone with palaeontological experience be sent to Ireland. He requested that Baily be transferred to the rank of Senior Geologist at a starting salary of £250 per annum. This was to rise annually by £10 until a maximum of £300 was reached (GSM 1/7, pp. 416–417).

This request was sanctioned and Baily joined the Irish branch as Senior Geologist (Acting Palaeontologist) in July 1857. He joined a Survey that contained men of some quality.

Baily's role in the GSI and the assembly of its fossil collections

Newly established in 1845, the GSI inherited the Portlock Collection and briefly occupied the Custom House, before moving to 51 St Stephen's Green in 1846. In 1870, it moved around the corner to 14 Hume Street where Baily occupied a top-floor room. The employment of fossil collectors meant that from the outset specimens were being accumulated, under the general guidance of the Director, and locally through the geologists themselves. Herries Davies (1983, 1995) has recorded in some depth the development of the Museum of Irish Industry under Sir Robert Kane, in an uneasy coexistence with the GSI. Within this museum, the GSI collections were kept and then displayed to the public, then later as the Museum of Science and Art, before a brief exposure within the National Museum of the fledgling Irish Free State. In historical analysis, it is simply a classic modern collection, very typical of many natural history collections that were critical to the establishment of numerous western European larger and national museums.

The collections themselves are the product of deliberate collecting in the field. Baily was not the sole collector based in the Dublin office of the Geological Survey the assembly of the fossil collection was necessarily a team effort, the long-term accumulation (1845–1890 in its main phase) of significant specimens by a small but dedicated group. Specimens were sent in by the geologists and the fossil collectors for examination by the Acting Palaeontologist. In 1858, Instructions to Officers engaged in the Collection and Determination of Fossils were issued from London, but they seem to have been largely stating and reinforcing what had already become practice, rather than establishing a new routine. Usually, the areas in which the Fossil Collectors should work were determined by Jukes, but on occasion when he was absent due to illness, Baily was instructed to direct the activities of the fossil collectors.

On some occasions, Baily would travel into the field himself or with the other geologists to sample productive horizons. The *Annual Report* for 1858, for example, noted that Baily spent twenty-eight days that year out collecting. His efforts seem to have been focused on the interesting sites, the topical or those critical to the understanding of regional stratigraphy. While in the field he did encounter some difficulties: in 1861, he was stopped excavating at Kiltorcan, Co. Kilkenny by the landowner (GSI letterbooks 26 August 1861). This site had produced wonderful Old Red Sandstone facies land plants including *Archaeopteris* and the freshwater mussel *Archanodon*, and was the subject of a number of papers by Baily (1859a, 1860a) In 1866, Baily and the two Fossil Collectors employed by the GSI were visited by the local police who though their activities were suspicious (GSI letterbooks 29 October 1866). Later, Baily requested the assistance of one of the Fossil Collectors and later still Alexander McHenry (1843–1919) was taken on as an Assistant Palaeontologist. Even in Dublin problems did arise—on one occasion he complained that coal was being stolen overnight from his room, inferring that it was too cold to work!

The Annual Reports give a clear record of the number of specimens, averaging 3,000–4,000 each year that had been collected. Of these, once identifications had been made and confirmed if necessary, specimens were selected for the Museum and stuck to labelled tablets, others selected for distribution to other branches of the Geological Survey in London and later Scotland, and some were rejected as worthless and thrown out. In 1868, Baily stated that he had examined over 30,000 fossils since his arrival in Ireland, had arranged 12,000 of these on labelled wooden tablets, and had distributed 5,000 duplicates (GSM1/16, p. 168). Increasing the total sum of knowledge of Irish stratigraphy was the purpose of collecting, not to simply increase numbers of the collection.

Another of Baily's main duties was to provide accounts of the fossils encountered during the mapping programme of the GSI. In total, Baily contributed reports on Palaeontology to sixty-six *Memoirs*. For many of these, he provided woodcuts illustrating the most important or stratigraphically interesting fossils. Jukes and others may have directed the mapping programme and possibly even the day-to-day work of the Fossil Collectors, yet ultimately Baily was the architect of meaning from the collection. In several letters within the letterbooks of the GSI, Jukes was at pains to insist that he alone, as Director, should have the final say on the stratigraphy, and what formation, in the assumption of direct equivalence of the geology found in Britain, different rocks belonged to, based principally on their fossil content. Baily was required to defer to his superior officer's judgement, yet realistically, Jukes had to rely on Baily for the identification and differentiation of different fossil species.

Baily's status within GSI: a cause for embitterment

Although initially sent to Ireland as a Senior Geologist which put him second equal in that salary scales for the Irish office after the Local Director Jukes, and equal with George Victor Du Noyer, Baily quickly began to seek further promotion, and thus an increase in his salary. He never attained the full title of Palaeontologist which he left would put him on equal rank with Salter and others in the GSGB, and in time, this rankled heavily with him. Whether the lack of title was the main source of upset, or whether it was his impecunious state was forefront of his pleading we shall never know, but both were factors in his unhappiness over a period of 31 years. He regularly sought promotion through letters to the Lords of the Department of Education, via the Local Director Jukes, and later Edward Hull. He submitted Memorials, detailed pleas for due recognition for his abilities, his responsibilities, long service, and the perceived injustice of his appointment not being made a full one.

He first applied for promotion in October 1859 only a short time after arriving in Dublin (GSM 1/8, p. 27), and when that failed he later applied for his post to be raised to Palaeontologist. This appeal was rejected by the authorities in London as they didn't believe that the Geological Survey needed two palaeontologists (GSM 1/8, pp. 321–322). By 1864, his salary was £340 and it reached a maximum of £350 per annum in 1865. On 1 April 1868, Baily was appointed Demonstrator in Palaeontology in the Royal College of Science for Ireland, a position that was in addition to his post at the GSI, and for this, he received an additional £100 per annum (GSM 1/9, pp. 21–22). In recommending him, Murchison wrote:

> Mr W.H. Baily is a most-fit and proper person to fill the new office; and being a deserving public servant, with a large family, it has long been my wish to see him recompensed. Mr Baily is really a sound Palaeontologist
> (GSM 1/9, p. 22)

For this, he had to give a number of classes in Dublin and in other towns in Ireland. He held this position until his death.

In July 1867, Baily wrote to Murchison asking that his status be put on a level with the District Surveyors who received £500 per annum (GSL archive LDGSL/838/B/4) (Williams, 2017) but his request fell on deaf ears. In June 1868, he requested a rise given he had to pay a premium of £50 per annum in life insurance and had also to pay income tax. He felt that given the nature of his job, and the fact that he had a large young family that he needed insurance cover in case he had an accident and couldn't provide for them (GSM 1/16, p. 168). He also pointed out that the move to Ireland cost him personally £100, and that due to the high cost of living, clothing, provisions and house rent (GSI letterbooks 10 February 1866) that he had to sell his London home to cover expenses (GSI incoming Guardbook 9 September 1872). Sadly for him all his efforts and pleas fell on deaf ears, with senior officers refusing on occasion to even submit his applications so soon after a previous refusal, for fear of displeasing 'the Lords'. Some correspondence between Murchison and others points to the fact that his salary was not disparate from that of the 'Palaeontologist' in the GSGB (GSM 1/8, p. 480; 1/9, p. 251)—Salter had started on a salary of £250 when appointed in 1854 (Secord, 1995, p. 66), and that in any case, the responsibilities in London were greater than in Dublin, although Salter's role there was soon reduced due to his various limitations (Secord, 1995, p. 66), and that in the wider view from London, his claims were unsubstantiated.

He applied for the vacant Chair of Zoology at the Royal College of Science for Ireland, that had become available on the death of John R. Kinahan (brother of his colleague George), but although getting permission from Murchison to hold this position concurrently with his Survey post (GSI letterbooks 21 February 1863), his application was unsuccessful. It shows the extent that Baily would go in seeking additional funds by applying for a Chair that he frankly was poorly equipped to hold. Finally, in 1881, in his final attempt to free himself of his Irish post and its salary he applied for the position of Palaeontologist at the Survey in London (GSM 1/26, p. 100). This had become vacant when Robert Etheridge resigned, but by then, Baily was in his mid-60s and he was not appointed.

The main reason for these pleas was probably not simply a quest for additional status, but for the additional salary that would come with such promotion. Baily appeared to be frequently short of money, as Sir Roderick Murchison attested to when he received the Wollaston Fund for 1867 from the Geological Society of London on his behalf. Baily had been unable to receive the award in person from the President William Warrington Smyth, as perhaps, ironically, he could not afford to travel across the Irish Sea. Murchison commented: 'while my deserving friend Mr. Baily is rich in fossil lore, and, I may add, in children, he is, like too many men of science, poor in sovereigns' (Murchison in Smyth, 1868).

Throughout his career, there is evidence that he sought ways of augmenting his salary. Commissions for artwork certainly helped his financial state at times. Jukes who was also Professor of Geology at the Royal College of Science commissioned Baily to produce 48 large geological illustrations for use in his classes. For these, Jukes agreed to pay Baily £1 each. However, Baily was slow to deliver and even requested an advance of the fee. This later caused animosity and problems between the two men (see below). In 1862, Jukes published his *Student Manual of Geology* which contained many woodcuts of fossils executed by Baily (Figure 14.2). It is obvious that he was paid for such a commission, and we would suggest would not have undertaken it unless this had been agreed. Jukes wrote in the forward:

> By the liberality of the publishers I have been enabled to take advantage of the presence of Mr W.H. Baily in Dublin, who compiled for me lists of characteristic fossils, which, with some modifications, are those given in the third part of the work. Mr. Baily also drew on the wood the figures which make the fifty "fossil groups" by which that part is illustrated.
>
> (Jukes, 1862)

At times issues of finance, family, status, and the development of a scientific reputation threatened to overwhelm Baily and he was granted sick leave on more than one occasion due to anxiety and overwork. An uneasy relationship developed with Jukes, his immediate superior, and this finally exploded in August 1861 following a meeting between the two that initially took place in Baily's office but then reignited shortly afterwards in Jukes'. After the meeting, the ultimate sanction, that of the removal of Baily from Dublin was requested. On 27 August an exasperated Jukes wrote to Murchison in London and stated that

> I can no longer tolerate Mr Bailys behaviour here…I must therefore request that Mr. Baily be removed from the Irish branch of the Geological Survey at the close of this official year or as much earlier as circumstances will allow.
>
> (GSI letterbooks 27 August 1861)

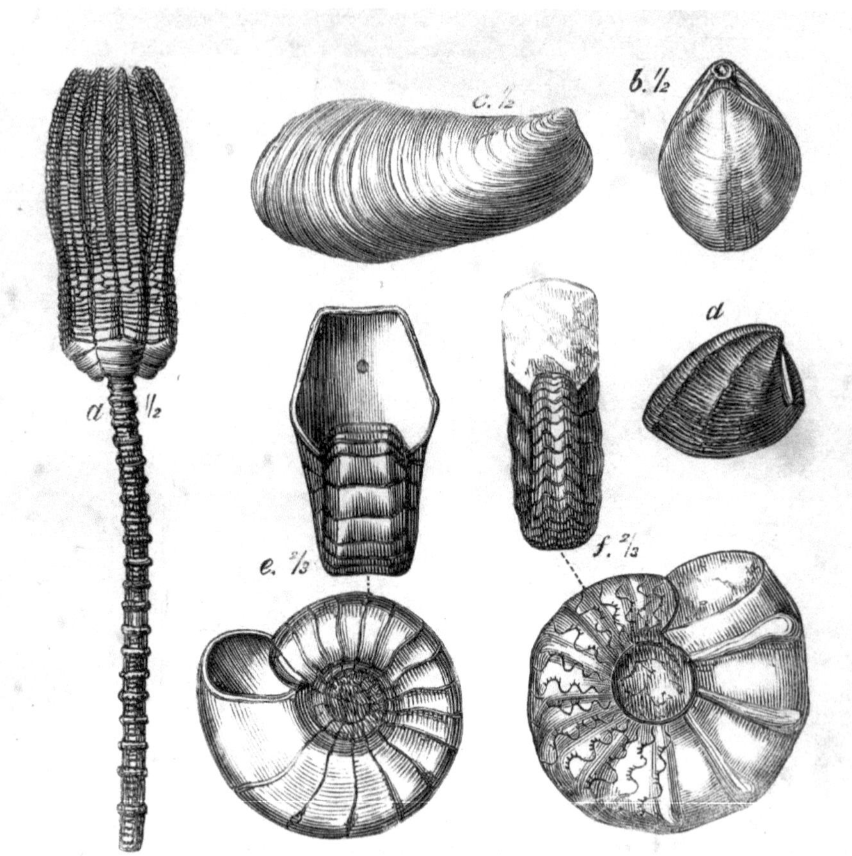

Figure 14.2 Fossil Group 22. Muschelkalk fossils. Woodcut by W.H. Baily used in J.B. Jukes *School Manual of Geology*, 1862 (Author's collection).

Almost immediately Baily stated that as he was on leave when the incident took place, it was a private matter and not subject to official record. He refused to apologise to Jukes as he felt that it was he who deserved an apology on account of being called an 'ungrateful beast' and 'damned scoundrel' (GSI letterbooks 14 October 1861).

A month later Jukes followed his first letter up with a long missive to Murchison in which he gave his reasons for the request, and they give some insights into Baily's character (GSI letterbooks 26 September 1861). Jukes pointed out that this had not been the first incident but that he had held back making any complain on account of his concern for his Senior Geologist's family who would be destitute should he lose his position, and the fact that he was not keen to lose the palaeontological expertise that Baily possessed. He complained that *Memoirs* had been delayed as the palaeontological report was not submitted to deadline. He pointed out that on Baily's arrival in Dublin he had been told that there was a great deal of work to be done sorting out the collections Jukes considered that little work had been done to remedy the chaos, and when he mentioned this, Baily 'burst into a fit of fury, and with black looks, a loud tone, and insulting gestures, told me that he had been longer on the Survey than I had.' Jukes also lamented paying Baily in advance for some commissioned artwork, and this incident was fuelled further when Baily requested an advance on additions work so that he could attend the British Association meeting in Manchester where he could meet his friends, relax and present a paper. Jukes refused to advance the money and suggested that Baily travel down the country and spend a leisurely week collecting fossils; Baily retorted that fieldwork actually cost him money. Jukes replied that in any case if Baily was short of money spending it attending a scientific meeting didn't make sense. Baily felt that participation at meetings would raise his scientific reputation; Jukes thought otherwise.

Jukes did acknowledge that his inferior officer had in the recent past had got through a great deal of but 'it has been done in a grudging and reluctant manner, as if he were working under an oppressive taskmaster, and he has never given me that hearty and cheerful assistance, I had a right to expect of him.'

By 23 October, Baily had simmered down and thought it prudent to request Murchison to disregard his letter of 14 October 1861, and the following day he relented and said that he was willing to make the 'requisite apology to Mr Jukes' (GSI letterbooks 24 October 1861). No doubt Murchison had set out his stall, quenched the flames of argument, and told the protagonists to behave appropriately.

Baily wasn't removed from his Survey position, and even after the events of August 1861, his lack of promotion seems to have been a thorny issue that under-pinned much of his activities. In all probability, it accounts for a degree of tardiness in producing reports for the Memoirs, in a reduction in the rate of fossils examined, and a withdrawal from the affairs of geologists in Dublin, more evident in his later years. It is hard not to feel some sympathy for the injustice of his career stasis; however, much of it was perceived rather than real.

Baily's involvement in scientific circles in Dublin

Dublin was a thriving city in the mid-1850s and had recovered from the loss of its Parliament and direct rule in 1801 to be regarded as many as being the second city of the Empire. Many institutions catered for the scientific populace, of which the Royal Irish Academy and the Royal Dublin Society were the most prestigious. Baily became involved with both and with the smaller specialist geological community in Ireland.

On 9 December 1857 soon after arriving in Dublin, Baily was proposed for Fellowship of the Geological Society of Dublin; his proposer was Jukes, his new boss, and his seconder Du Noyer, his GSI colleague. The nomination accepted, Baily paid soon after the entrance fee of £1 and the Annual Subscription also £1. For the remainder of his life, he was involved with the organisation and with its successor the Royal Geological Society of Ireland, but it was a relationship that was not always convivial. The Society met monthly for General Meetings when up to four papers were read, some of which were later published in its *Journal*. Baily attended his first meeting in March 1858, and read his first paper 'On *Limulus* from the Coal Measures Bilboa Collery, Co. Carlow' at the 9 June 1858 meeting. He served on its Council for several periods (1859, 1863–1865 and 1871–1874) and published seventeen papers in its *Journal*, and the Minute Books and records of meetings now housed in the Geological Museum, Trinity College Dublin and where the preceding information is drawn, also show that he was actively involved in many meetings and frequently spoke and contributed opinions and judgements on other people's papers, as well as presenting his own.

However, he did not maintain sustained and unbroken Fellowship. On at least two occasions, he resigned his Fellowship: in 1865, after when he rejoined again two years, and again in 1874 at which time he also resigned his place on the Council. The reasons for the second resignation lie in fees payable for the production of illustrative plates. In the Minutes of the Council Meeting, it was 'Resolved that the remuneration to be paid to Mr Baily for illustrating his own papers be at the rate of one guinea for each plate.' (RGSI Minutes of Council Book). A cheque was authorised in favour of Baily for two guineas. One can only imagine that Baily wanted to be paid to prepare his own illustrations for his own papers and that the Council either resisted such a move or wished to pay less than was demanded. Following some accommodation on the matter, and agreement on payment, it is noted at the bottom of the same Council Meeting minutes that he had resigned from the Council. Following his second resignation, Baily failed to rejoin the Society even though he continued to publish papers in its *Journal* and attend some monthly meetings as a Visitor often in the company of his son H.W. Baily (RGSI Meeting Book).

He also contributed a small number of scientific communications to the Royal Dublin Society whose headquarters were in Leinster House, an impressive seventeenth century townhouse which now serves as the Dáil (Parliament). Perhaps, Baily's most significant contribution to this Society was his publication in 1863 with Alexander Carte (1805–1881) of *Plesiosaurus cramptoni* (now *Rhomaleosaurus cramptoni*) a large plesiosaur, which they named for the famed Dublin surgeon Sir Philip Crampton (Carte and Baily, 1863; Benton and Taylor, 1983, p. 415). This specimen had been collected in 1848 from the Upper Lias at Whitby in Yorkshire and moved to the home of the Marquis of Normanby. Later, it was given to Crampton who moved it to Dublin and donated it to the Royal Dublin Society. They then displayed it in their museum, and Carte and Baily produced a detailed description with a fine lithograph by the latter. Today, the specimen is not on display in the National Museum of Ireland into which the Royal Dublin Society Museum became in late 1877, but recently its impressive skull has been prepared (Smith and Dyke, 2008).

He was also active as an Honorary Member in the Dublin University Zoological and Botanical Association which had been established by the naturalist Robert

Ball. Designed to draw together the academics, professions, and students under one umbrella organisation, Baily contributed a number of papers which were published, as well as serving on its Council for a period.

He was elected a Fellow of the Linnean Society on 19 March 1863 but it appears that Baily made few contributions to the life of the Linnean Society. Nine years later on 8 April 1872, he was elected a Member of the Royal Irish Academy. This was an honour of some significance as not all men of science at the time were admitted to Membership. On 9 February 1873, he read to the Academy a paper on the Kiltorcan fossil flora which was subsequently published in its *Proceedings* (Baily, 1877) and in the same serial published jointly with George Kinahan on rocks in Co. Tyrone (Kinahan and Baily, 1880).

He was also an Honorary Corresponding Member of the Royal Society of Liège, of the Natural History Societies of Dresden and Strasbourg, and of the Geological Society of Belgium (GSM1/26, p. 101).

Baily and the BAAS

Like many Irish-based scientists, Baily participated in many of the BAAS meetings that were held annually in towns and cities in Britain and Ireland. He joined in 1857 but made his first contribution, on fossils from South Africa, earlier in 1854 at the meeting in Liverpool. In all, he made twenty-two contributions and in general these, with the exception of Committee reports were published in the Annual Reports as abstracts and full papers often appeared elsewhere. He enjoyed attending the meetings where he could meet his friends but he also felt that they offered him the change to enhance his scientific reputation (GSI letterbook Jukes to Murchison, 27 September 1861). Between 1856 and 1883, he missed the Oxford meeting in 1860 probably on account of the tragic loss of two of his children that year and didn't attend between 1865–1867, 1870–1874, and 1876–1877. He must have made a special effort to attend the meeting of 1875, and no doubt participating gave him great pleasure as it was held in Bristol, his home town.

Much of the work of the BAAS was carried out by Committees appointed to investigate various topical issues. Baily sat on one and acted as Reporter (later termed Secretary) to these: the Committee that studied the fossil flora at Kiltorcan (1869) for which the Association awarded £40 (BAAS Annual Report for 1869, 1870; Howarth, 1931), the Committee that studied the fossil flora of Ireland (1873; £20) and the Committee appointed to study and collect the Tertiary Flora of the North of Ireland (1880–1882; £20). He wrote on the Kiltorcan flora at the Norwich meeting in 1868 (Baily, 1869) and reported on behalf of the Committee at the next meeting in Exeter (Baily, 1870). No report was forthcoming for the Committee on the Fossil Flora of Ireland established at the Brighton meeting in 1872. However carried out a great deal of work on the Tertiary floras of the north of Ireland and penned four reports for the Association in the 1880s. Three of these were illustrated with lithographs by Baily.

How the money granted by the BASS was spent by the committees is unknown but given the reports bear all the hallmarks of Baily's work, and show little evidence of input from colleagues it is reasonable to suggest that Baily received most of it to aid his research.

During Baily's career in Ireland, the Association met on the island on three occasions (Morrell, 1990). The first was in 1857 in Dublin and the second was in 1874 in Belfast. This prompted Baily to write a short sketch of the geology of the city and its neighbourhood (Baily, 1874). Four years later, the Association met in Dublin and Baily served on the committee charged with producing the official handbook. For this, he penned an account of the palaeontology of County Dublin (Baily, 1878a) and this contribution was also published elsewhere in 1880 in both the *Journal of the Royal Geological Society of Ireland* and the *Scientific Proceedings of the Royal Irish Academy*. As for the earlier Belfast meeting, he wrote a short account of the geology of the area adjacent to the host city (Dublin and Wicklow) which similarly appeared in *Hardwicke's Science Gossip* (Baily, 1878b).

Baily's research and published works

Amongst the numerous papers Baily had published, the most noticeable feature is the diversity of topics which Baily had sufficient grasp of to expound on in print. Coupled with a natural talent, his early experience in Bristol and London illustrating fossils for other people, and then across the gamut of Irish geology, was more than adequate for him to tackle the full spectrum of fossil animal and plant kingdoms, and the entire Phanerozoic. Although he was competent with fossils, his naturalist abilities with modern species were more limited, and lists of shells from ice-dredged marine deposits around Ireland were often provided by other specialists for the GSI memoirs. Indeed, he was not unaware of living relatives of the fossils he identified.

In many instances, the same paper appeared in more than one journal, for example his paper on *Solarium* published by the Dublin University Zoological and Botanical Association (Baily, 1860b) was republished in the *Natural History Review* edited by the Trinity geologist Rev. Samuel Haughton, and it reappeared a third time in the *Annals and Magazine of Natural History*. This was a normal practice in Victorian scientific circles and not an anomalous circumstance associated with Baily himself.

In general, his research interests can be grouped into several major categories: fossil chitons and arthropods, fossil plants of various ages, and fossil amphibians, although his papers did cover other areas too.

He became interested in fossil chitons, members of the molluscan class Polyplacophora following a discovery of specimens in Carboniferous limestones in Limerick (Baily, 1859b). He published a number of papers in 1860, one of which discussed possible Silurian chitons. Only recently has Baily's work been reassessed using his original material from the GSI collections and his taxon *Chiton thomondiensis* has been reassigned to *Pterochiton thomondiensis* (Sigwart, 2007).

He developed an interest in palaeobotany and reported the taxon *Sphenopteris hookeri* for the first time from Kiltorcan (Baily, 1860a), and another plant *Sagenaria bailyana* was named for him by Wilhelm Philipp Schimper. He also worked on plants and amphibians from the Jarrow Colliery (Baily, 1866, 1876) and reported the presence of the rare crustacean *Belinurus* from central Ireland (Baily, 1862).

His final enterprise was a small book entitled *Rambles on the Irish Coast* in which he described the geology, natural history, and antiquities of the Dublin coastline. It was his intention to 'issue a series of the "Rambles" to be published occasionally, each being complete in itself, and the parts so arranged as to form, topographically, a continuous work' (Baily, 1887). Death intervened before he could issue more than one volume.

Figures of Characteristic British Fossils: his masterwork

Baily had a vision of a gap in the market, and with his aspirations for more income, in the 1860s, he set about producing an affordable identification guide to common fossils for the average person, or at least for the non-specialist palaeontologist. It was certainly not an original idea, and in fact may have arisen from his earlier working relationship with Forbes, the Geological Survey Palaeontologist. Entitled *Figures of Characteristic British Fossils: with Descriptive Remarks, volume 1, Palaeozoic* (1867–1875), publication commenced from 1867 onwards in four small sections of about ten plates each issued at a cost of five shillings. When in 1875 all the Palaeozoic periods had been covered providing 700 fossil illustrations in monochrome or watercoloured (some of which were executed by his daughter at home (Williams, 2017, p. 466), a title page was issued (in at least two different styles) (Figure 14.3) for the set to be bound together with an introductory section comprising the Descriptive Remarks.

Unfortunately, despite the award of the Wollaston Fund from the Geological Society of London in 1867 towards the production of plates for the book (Smyth, 1868), the project was not to have been the financial success that Baily must have hoped for, and no further volumes were produced.

Family life in Dublin

In 1847, Baily married Ann Elizabeth Rawlins (1822–d. 17 February 1887) (Letter J.W. Salter to De la Beche, 22 October 1847, NMW De la Beche archive 1853), and the couple had eleven children, at least three of whom died young (Further details of his family and Dublin addresses are given in Wyse Jackson and Parkes, 2009, pp. 75–76). A family of this large size was not unusual in Ireland at that time, even for Protestants such as the Baily family, and supporting this number must have been difficult for most of the time. He often cited family illnesses as reasons for his seeking promotion. He himself requested and was granted sick leave in June 1865 on account of suffering from the symptoms of 'lead disease' as

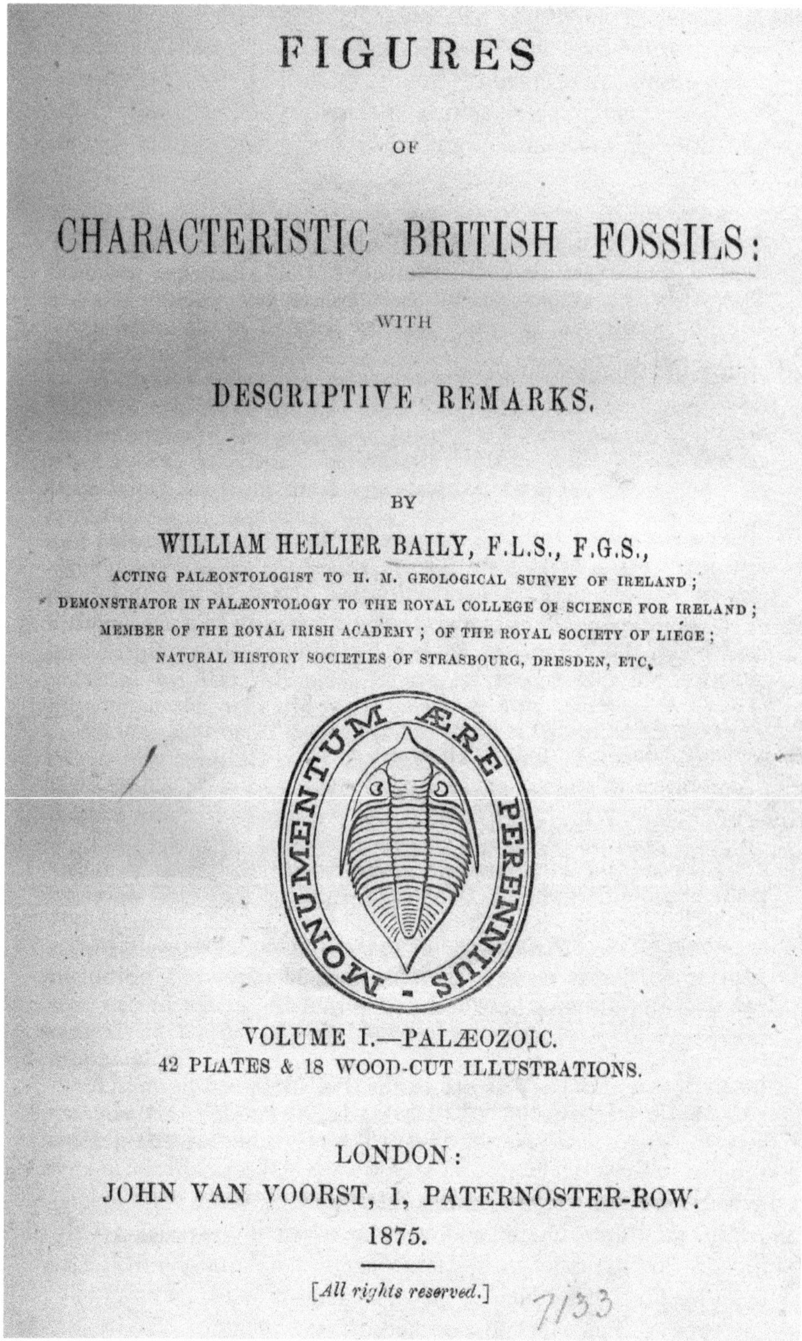
FIGURES

OF

CHARACTERISTIC BRITISH FOSSILS:

WITH

DESCRIPTIVE REMARKS.

BY

WILLIAM HELLIER BAILY, F.L.S., F.G.S.,

ACTING PALÆONTOLOGIST TO H. M. GEOLOGICAL SURVEY OF IRELAND;
DEMONSTRATOR IN PALÆONTOLOGY TO THE ROYAL COLLEGE OF SCIENCE FOR IRELAND;
MEMBER OF THE ROYAL IRISH ACADEMY; OF THE ROYAL SOCIETY OF LIÈGE;
NATURAL HISTORY SOCIETIES OF STRASBOURG, DRESDEN, ETC.

VOLUME I.—PALÆOZOIC.
42 PLATES & 18 WOOD-CUT ILLUSTRATIONS.

LONDON:
JOHN VAN VOORST, 1, PATERNOSTER-ROW.
1875.

[*All rights reserved.*]

7133

Figure 14.3 Title page of *Figures of Characteristic British Fossils with Descriptive Remarks*. Volume 1. 1875 (Author's collection).

his doctor attested (GSI letterbooks, 28 June 1865). Could this have been a direct result of his long career as a draftsman and palaeontologists?

Baily and his family lived in a variety of locations in the Dublin suburbs of Ranelagh and Rathmines. These were among the several new townships built outside the city limits which were granted the privilege of local self-governance (Ó Maithú, 2003). Thus, the area became popular with the professional middle classes of which Baily in his position would have regarded himself as being a part. The houses were well-constructed, large and generally were accompanied by a garden, which was in marked contrast with much of the property in the city. The townships were also easily accessible to Baily's place of work, being well served by trams and later a suburban rail line. Baily and his family moved fairly frequently and this suggests that they rented rather than owned property.

Baily's death

Baily requested a period of three months of sick leave on 8 March 1888 on account of threatened dropsical attacks, known today as oedema, an accumulation of fluids in body organs. In June, 1888 he applied for an extension of another three months, and again the leave was granted but in doing so the Director General of the Geological Survey Archibald Geikie wondered whether the time had come for retirement to be considered. Baily never retired but died in service on 6 August 1888 when aged 69.

He was buried in Mount Jerome Cemetery in Harold's Cross, a suburb of Dublin adjacent to Rathmines. His wife who pre-deceased him a year earlier in February 1887 shares his grave in Grant (Plot) 2654 in Sector 187/B1 and that of three of their children. The grave is marked by a Portland Stone headstone (Wyse Jackson and Parkes, 2017).

What is surprising is that within the GSI his passing was barely commented on. No obituary is even bound into the GSI Guardbooks as it was for Jukes, even though a number were published. Although 'lamented' by Hull and others, Baily's passing seems to have been quietly and without fuss. An estimate for the year 1890 in the letterbooks has a poignant blank space in the staff complement where Baily has been ghosted out. The GSI mapping was completed around this time, and the *Annual Report* as Director for 1888 noted:

> This memoir when published will make the numbers of those to which Mr Baily contributed Palaeontological Notes no fewer than 66. Little now remains to be done in Irish palaeontology except to add to the collections from time to time from new openings.

With this, Edward Hull sounded a death knell for the palaeontological tradition of which Baily had been a key element for 44 years. Baily, the Acting-Palaeontologist was replaced by a petrologist John Shearson Hyland (Herries Davies, 1995, p. 105), perhaps given that his field was covered Alexander McHenry who had served as Acting Assistant Palaeontologist to Baily for eight years (Herries Davies, 1985, p. 315).

Vale

In the obituary published in the *Geological Magazine*, Baily is described as belonging to:

> that small body of Geologists Palaeontologists who possessed an extensive general knowledge both of rocks and fossils, and also the invaluable ability to draw, as well as to describe, what they saw and studied, whether in the field or in the cabinet. These men can never be replaced by our modern student-specialists.

The author of the obituary above went on to state that 'Mr. Baily was of a genial disposition', although given what Jukes said of him it is clear that he could be temperamental at times. While the problems that he encountered regarding his state, status, and salary while employed in the service of the GSI may have caused him to be fractious and difficult at times, we feel some sympathy for his position in his never being accorded the title 'Palaeontologist' having to endure the indignity of being the 'Acting-Palaeontologist' for all of his Irish career. However, it is clear that considerable efforts were made by his colleagues to increase his salary by various means and that by 1868 he commanded an official income of £450 which was close to that of the Local Director of the Irish Branch of the Geological Survey.

Baily was a good curator of the palaeontological collection in the GSI and was the reliable hand responsible for their original documentation. These specimens have proved invaluable over the years providing biostratigraphic context of Ireland's geological history. Equally, Baily's other scientific legacy lies in his palaeontological contributions, written and drawn, and his beautiful book *Figures of Characteristic British Fossils*.

15

WILLIAM THOMSON'S DETERMINATIONS OF THE AGE OF THE EARTH

In R. Flood, M. McCartney and A. Whitaker (eds) *Kelvin, Life, Labours and Legacy.* Oxford University Press, 2008, pp. 160–174; 332–334.

Introduction

For four decades in the nineteenth century, William Thomson held the centre stage in the debate on the age of the Earth (Brush, 1982; Richter, 1986; Burchfield, 1990; Stinner, 2002). His domination of this subject began in the 1860s and lasted largely unchallenged until the close of that century, and he was pivotal in promoting the viewpoint of the physicists that often ran contrary to the opinions held by the geologists.

By the mid-1800s, it was widely recognised that the Earth was much older than the earlier studies of both the biblical texts and pseudoscientific methodologies had suggested. Gone were the chronologies of Archbishop James Ussher (1581–1656) and countless others who had scanned the pages of the Bible for the answer to this question. Confined to the pages of fantasy were the thoughts of Edmond Halley (1656–1742), the astronomer who in 1715 had suggested a method that examined the saltiness of the ocean to date the Earth. Ignored too was the method of the Oxford academic and Keeper of the Ashmolean Museum, Edward Lhwyd (1660–1709) who had suggested that counting the number of boulders lying on the floor of the Vale of Llanberis could point to a similar conclusion (Wyse Jackson, 2006a).

Geology as a discipline was firmly established in the early 1800s. The Geological Society of London was founded in 1807, universities began to establish professorships in the subject, and in 1845, the Geological Survey of Great Britain and Ireland commenced work. By this time, most practitioners of this emerging science realised that the Earth was very ancient, but few could say how old it was in absolute figures. The Scottish geologist-gentleman farmer James Hutton (1726–1797) had in 1785 published an abstract of his theory of the Earth in which he argued that the Earth's history was cyclical and basically of infinite duration.

 DOI: 10.4324/9781003475552-18

Charles Lyell (1795–1875) who, through his popular geological books had brought the new subject to a huge audience, had championed the idea that the processes acting on the Earth at present were similar to the processes that had acted on it in the past. Change was slow, and uniform; this conclusion was in marked contrast to the earlier ideas in which events were sudden or catastrophic. By the 1840s, most geologists had accepted Lyell's ideas on the uniformitarian nature of Earth's history, and adhered to the doctrine that the Earth's history was infinite.

Geologists had, by the mid-nineteenth century, also established a well-defined stratigraphical order of the rocks exposed at the Earth's surface. On the basis of geological mapping, the distribution of various geological horizons was well known and in Britain it had been pointed out by William Smith (1769–1839), the surveyor and author of a celebrated geological map of England and Wales, that each rock unit could be recognised and identified through examination of its fossil content, and that those found at the top of a sequence were younger than those found at its base (unless of course there was clear evidence that the whole sequence had been overturned). This allowed geologists such as the Cambridge professor Adam Sedgwick (1785–1873) and the soldier-turned-geologist Sir Roderick Impey Murchison (1792–1871), and others, to define various geological periods each of which contained rocks with particular physical characteristics and fossils. These periods make up what is termed the 'geological column' and by 1860, the relative positions of one period to another had been worked out. However, the absolute ages of the boundaries of the geological periods, and their duration, were not known and would remain unresolved until the 1920s.

Palaeontology, the study of fossils, played an important role in the development of the geological column. Today, it is largely forgotten that Charles Darwin (1809–1882) began life as a medical student and that it was when he was unable to continue his studies in Edinburgh that he turned his attention to geology. Prior to sailing on H.M.S. *Beagle*, he undertook fieldwork in north Wales, and many of his observations made during his epic journey were geological in nature. He published on volcanoes and coral reef islands. On his return to England, he turned his attention to biological evolution, but in *The Origin of Species*, published in 1859, he also wrote on geological time; he argued that the topography of the Weald in south-east England had taken 306 million years to form. His theory of evolution had many sceptics because it appeared to them that there was not enough time available for biological evolution to have taken place, yet through his Wealden example Darwin had provided enough time. His ideas brought him into conflict with many thinkers and men of science who fundamentally could not accept his ideas on biological evolution. One of these was John Phillips (1800–1874), successively Professor of Geology in London, Dublin, and Oxford, who in 1861 examined the rate of deposition of sediments and argued that the pile of rocks that constituted the rock succession could only have taken 54 million years to form, and that Darwin's estimate based on the denudation of the Weald was very wide of the mark.

Darwin also came into conflict with William Thomson, the Professor of Natural Philosophy in the University of Glasgow. However, as Stephen Brush put it in 1982, the rationale that the argument between Thomson and the geologists and Darwin was about evolution was, and continues to be, overstated (Brush, 1982). Thomson's main bone of contention with Darwin's logic was that it didn't follow the laws of physics. That lay at the root of the debate between the physicists and the geologists. He didn't take a stance against biological evolution: he even went so far as to suggest that it could proceed at a faster rate than imagined by the biologists, and given this his shorter timescale would not be problematic.

Thomson on the age of the Earth: a three-pronged attack

Thomson was infuriated by Charles Darwin's dabbling in geochronology and was also not impressed by writings of the 'modern geologists' as he termed those who followed the uniformitarian scheme, and so he set out to prove the actual age of the Earth through the application of the laws of physics. Early on, while formulating his ideas, Thomson had the sense to seek out some geological advice and discussed the validity of Darwin's ideas with none other than John Phillips. Given Phillips' stance against Darwin, Thomson wittingly or unwittingly had acquired an ally within the ranks of the geologists. Nevertheless, Thomson's findings did not, by and large, satisfy the geologists, and thus begun a feud that lasted until the death of Thomson over 40 years later. This stand-off between the physicists and the geologists continued until the mid-1920s when the latter were arguing that the Earth was older than the age suggested by the former for the universe.

Thomson applied himself to determining the age of the Earth in three ways: the first was in relation to the Sun, and he attempted to estimate how long it had been shining and used this as a corollary for the age of the Earth. Second, Thomson investigated the effect that friction caused by tides might have had on the shape of the Earth. The third method took the secular cooling rate of the Earth, and it is for this work that Thomson is chiefly remembered in the geochronological field. He revisited the research of George-Louis Leclerc, Comte de Buffon (1707–1788) nearly a century earlier. Buffon, a French aristocrat had been Curator of the celebrated Jardin du Roi (now the Jardin des Plantes) in Paris and had built a forge on his estate near Dijon, where he heated spheres of different materials and sizes and measured their cooling rates. From this, he estimated that the Earth was nearly 75,000 years old (Wyse Jackson, 2006a). This figure was considerably higher than that thought correct by the bulk of Buffon's contemporaries, Thomson would arrive at a figure that was higher still.

Whether by design or coincidence, these very different schemes worked-up by Thomson ultimately came to broadly the same conclusion in terms of age determination, and this led to his unwavering belief that his determination was correct. It led too to considerable support from colleagues and was partially instrumental in his pre-eminence in British science for such a long time.

The age of the Sun and the origin of its heat

Thomson produced a large volume of work on the Sun and its heat, and first published on the subject in a number of papers published in the *Transactions of the Royal Society of Edinburgh* and in the *Philosophical Magazine* in 1854. Drawing on the work of James Prescott Joule (1818–1889) published ten years earlier, who demonstrated the relationship between energy, heat, and force, Thomson discussed the origins of the mechanically derived heat of the Sun. Energy, lost as radiant heat and light could not, he noted, ever be replenished. Noting some earlier ideas that accounted for the source of the Sun's heat, Thomson rejected out-of-hand the notion that the Sun was a heated body, one assumes with its own source of heat. He considered that the Sun formed due to the attraction meteorites which built up its mass. As they were pulled towards the Sun, they burnt up and the frictions caused by the vaporised gases were released as heat. In this, he was following the scheme suggested by John James Waterston (1811–1883) the previous year. While the bulk of this paper was theoretical he published only one quantitative calculation. He argued that it was possible to determine how long it would have taken for the meteorite component of the Sun's mass to accumulate. This was only a proportion of the Sun's bulk. This he said took 32,000 years and suggested that the Sun would not have shone on the Earth for many more times than 32,000 years. He remarked:

> We may be confident, then, that the gradual augmentation of the Sun's bulk required by the meteoric theory to account for this heat, may have been going on in time past during the whole of the human race
>
> (Thomson, 1854; reprinted 1884, p. 8)

In terms of geological time, he pointed out that the Sun could have taken two million years to grow, which in his opinion was an adequate length of time required for the geological processes that had been described by leading scientists such as Charles Lyell.

Initially, the surface of the Earth, however, would have remained in a molten state due to the gravitational energy produced when meteorites collided with it. Following this period, the surface crust developed but it would have been too hot to allow vegetation to cover it.

Computing the amount of heat lost from the Sun to be 2,781 thermal units per second per square foot, which would equate to removing a layer fifty-five miles thick per year, Thomson rationalised that the Sun would burn out in 8,000 years time, and conversely that it must have been eight times its mass 8,000 years earlier. He went on to state:

> We may quite safely conclude then that the Sun does not get its heat by chemical action among particles of matter primitively belonging to his own mass, and we must therefore look to the meteoritic theory for fuel …
>
> (Thomson, 1854; reprinted 1884, p. 11)

Eight years later, Thomson (1862) returned to the question of the age of the Sun in a three-part paper published in *Macmillan's Magazine*. His earlier premise that meteorites falling into the Sun and the heat generated by friction of the vaporised gases as they burnt up simply was untenable. Meteorite falls would not have generated the volume of heat that he had suggested they would. Following the meteoritic theory of Hermann Ludwig Ferdinand von Helmholtz (1821–1894), Thomson modified his ideas on the role of meteorites in generating the Sun's heat, and rather than vaporisation being the source of the heat, it was the fact that as they moved the meteorites had a potential energy, which became kinetic energy as they moved towards each other, and which was converted into heat upon collision with themselves or the Sun. Consequently, he had to alter his time estimates and concluded:

> It seems, therefore on the whole most probable that the sun has not illuminated the earth for 100,000,000 years, and almost certain that he has not done so for 500,000,000 years
>
> (Thomson, 1862; reprinted 1889, p. 368)

However, he was more comfortable with the conclusion that the Sun had been operational for between 20 and 60 million years. Towards the end of his life, following persuasion by his associate Peter Guthrie Tait (1831–1901), Thomson favoured the lower age limit of 20 million years. This by association gave a limit to the age of the Earth.

Tidal friction

Immanuel Kant (1724–1804), the German philosopher and Professor of Mathematics at the University of Königsberg, realised that the Earth was slowing down due to the effects of tidal friction. Running with this idea, Sir George Howard Darwin (1845–1912), son of Charles, wrote extensively on the orbit of the Moon and the Earth's retardation rates. Some of his contemporaries even went so far as to hypothesise that the Moon had been formed by metastasis from the Earth. Terrestrial material was lost from the surface due to rapid spinning of the parent body and consolidated into the lunar sphere—a theory now known to be incorrect. Darwin calculated the time taken for the Earth and Moon to settle down from this initial rupture to their present condition and came up with a minimum of 56 million years.

In 1868, Thomson examined the Earth's history and chronology by focusing on changes in the Earth's shape over time, changes which were caused by a slowing of its rate of rotation over time. The rate he calculated was 3.6 seconds per annum. It is well known that due to friction on the surface waters as the Earth spins, tidal waters tend to become banked up and do not act in the predictable manner normally expected. The Earth, Thomson said, assumed its flattened spherical shape soon after its formation, while it was still molten. He realised that if one took the present rotation rate of the Earth and used this to calculate what the shape of the globe would have been if this had been the primordial spinning rate, one would expect a spheroidally flattened globe of a particular shape. This expected shape he found was appreciably no different from the actual shape of the globe, and so he deduced that very little time had elapsed since the formation of our planet. He acknowledged that there were difficulties of actually determining accurately the parameters that fed into this methodology but was confident that the Earth was no more than 1,000 million years old, and that 100 million years old was the more plausible figure.

Secular cooling of the Earth

In his research into the cooling rates of the Earth, Thomson was greatly influenced by the work of French physicist Jean Baptiste Joseph Fourier (1768–1830), which he had studied when aged only 16. Fourier had written that the source of the Earth's heat was threefold: from primitive internal heat, from heating by the Sun, and from heat in the universe but had also suggested that the Earth was cooling down by conduction and provided mathematical models to quantify this.

Thomson believed that the primordial Earth was molten throughout, but that it developed an outer solid crust soon after its formation. He believed that it then developed a solid nucleus surrounded by molten rock, and a solid thin crust, and that it continued to solidify from its centre outwards as the internal heat migrated through the rocks by conduction and heat was lost from its surface. Once solid, but still hot, it lost heat by conduction rather than convection. Thomson drew on experimental work undertaken between 1837 and 1842 by James David Forbes (1809–1868) that had revealed a temperature gradient at three stations near Edinburgh (the readings continued to be recorded at Calton Hill, Edinburgh for the next thirteen years until 1854). Forbes had placed thermometers at three, six, twelve, and twenty-four feet below the surface at each station.

In a paper published in 1861, Thomson discussed the findings and implications of this work at length and observed that the effects of diurnal changes in temperature become obsolete at depths of greater than three feet below the surface (Thomson, 1861). The earliest studies of temperatures beneath the Earth's surface had been carried out in Paris from 1671 (Buntebarth, 2002) and it was well known by the mid-1800s that temperatures increased as one went down into the Earth. To further research, the British Association for the Advancement of Science, at its 1867 meeting in Dundee, appointed a committee to carry out more work in this area. It met between 1868 and 1877 and was awarded the considerable sum of 330 guineas for this work. Geothermal gradients had previously been investigated by taking temperature readings from various depths in mines or from boreholes; however, this work was not very accurate until the invention in the 1830s of specialised thermometers designed for the task. Even by the 1860s, results were not conclusive, and it was believed that the temperature at the Earth's centre was approximately 3,600 degrees Fahrenheit which is about half of what we now consider the temperature to be.

In April 1862, Thomson read his first paper on the cooling of the Earth that tackled some of the problems raised in his mind by the geologists. This paper was published two years later in the *Transactions of the Royal Society of Edinburgh* (and summarised shortly afterwards in *The Times*) and began with a full-fronted attack on their uniformitarian philosophy and the need for an extended or indefinite Earth history:

> For eighteen years it has pressed on my mind, that essential principals of Thermo-dynamics have been overlooked by those geologists who uncompromisingly oppose all paroxysmal hypotheses, and maintain not only that we have examples now before us, that the earth, of all the different actions by which its crust has been modified in geological history, but that these actions have never, or have not on the whole, been more violent in past time than they are at present
>
> (Thomson, 1864; reprinted 1890, p. 295)

Thomson stated that the temperature increased 1°F every 50 feet downwards, and that this indicated that heat was being lost from the Earth's surface over a long period of time, and that the whole globe was cooling down. He rejected the hypothesis that the Earth had an internal, or as he put it 'temporary dynamical', or chemical source of heat. He argued that there was less volcanic activity at the time of writing that had been 1,000 years earlier, but that this fact had been ignored by most geologists. Less volcanic activity pointed to a cooler Earth. However, in the 1890 reprint of the 1862 paper he added, no doubt with some satisfaction, a footnote in which he said that most geologists in 1889 now believed the statement to be true! One wonders what Thomson made of the major volcanic eruption at Krakatau of 1883.

He carried out experiments to determine the conductivity of various rock types and took 7,000°F as the temperature of fusion of rocks, an estimate that had been determined a short time earlier. He argued, given this formula, that consolidation of the surface would have taken place 98 million years ago, or a maximum of 200 million years ago if one took 10,000°F (a high estimate, in his opinion) as the melting temperature of rock. Given the underground temperatures known, and the loss of heat due to conduction, he noted that if the Earth was only 40,000 years old then the temperature gradient would be 1 degree per foot. If 4 million years old it would be a tenth of a degree Fahrenheit per foot, and if 100 million years old it would be one-fiftieth of a degree Fahrenheit per foot. The latter figure is the temperature gradient figure that Thomson used and gave a coincident age with that of the consolidation age of the crust of approximately 100 million years.

Thomson also noted that it was possible, however, to generate the observed temperatures if the whole of the Earth's surface had been heated up to 100°F at some time in the last 20,000 years, but that this scenario was an impossibility as all traces of life would have been killed if the surface had attained this temperature. From examination of the cooling of the Earth, Thomson believed that was formed between 20 and 400 million years ago, with 98 million years being its likely age.

In arriving at this figure, Thomson made four assumptions about the conditions of the Earth. First, it was solid; second, it began at the same temperature throughout; third, that it must be homogeneous and have an identical conductivity throughout; and finally, no internal heat source was present. Subsequently, all of these assumptions were challenged and found to be untenable and incorrect. We know through the work of Richard Dixon Oldham (1858–1936) that the outer part of the Earth's core is liquid. The Earth is not homogeneous, and crustal rocks are quite different in mineralogy and density to those found in the Earth's mantle. Following the discovery of radioactivity by Antoine Henri Becquerel (1852–1908) in 1896, it was soon recognised that disintegration of radioactive elements provided an internal heat source for the Earth.

On 27 February 1868, Thomson revisited the subject of the Earth's antiquity when he addressed the members of the Geological Society of Glasgow. He used the occasion to try to bring the geologists around to his viewpoint and gave the assembled company a synopsis of the methods and results used in both the tidal friction and secular cooling schemes to age the Earth. He concluded that the Earth was no more than 100 million years old.

> … we find ourselves driven to the conclusion in every way, that the existing state of things on the earth, life on the earth, all geological history showing continuity of life, must be limited within some such period of past time…
>
> (Thomson, 1868, p. 25)

Detractors and supporters

Many of Thomson's detractors questioned his figures regarding the internal temperature and gradient in the Earth, and also questioned his reliance on a theory which they considered to be without any geological foundation. In 1867, during the British Association for the Advancement of Science meeting, Thomson fell into conversation with the geologist Andrew Crombie Ramsay (1814–1891) and the question of timescales arose (Thomson, 1889, p. 668). Thomson asked Ramsay:

> Do you think it [the Sun] has been shining on for a million million years?

to which Ramsay responded:

> I am as incapable of estimating and understanding the reasons which you physicists have for limiting geological time as you are incapable of understanding the geological reasons for our unlimited estimates.

Immediately Thomson retorted:

> You can understand physicists' reasoning perfectly if you give your mind to it.

Thomson was not the only one prone to caustic remarks. Thomas Mellard Reade (1832–1909), a civil engineer and geologist from Liverpool, wrote that 'Facts are safer than theories' (Reade, 1878, p. 154) implying that Thomson relied too heavily on the latter.

Thomson's ideas on a cooling Earth were at variance with the views of many other scientists, but not all. In 1865 the Rev. Samuel Haughton, (1821–1897) a Fellow of Trinity College, Dublin and its Professor of Geology, and Darwin critic, supported his fellow Irishman's work. Haughton reworked the calculations applied to the cooling method for age determination. In 1865, in his *Manual of Geology*, Haughton published the results of these calculations and which showed that the Earth was 2,298 million years old considerably higher than Thomson's findings. Haughton was no mathematical slouch and his calculations were accurate, but his problem was that the basic figures he used for the calculation were flawed.

Thomson's timescale was questioned in 1869 by none other than Thomas Henry Huxley (1825–1895) who asked in his Presidential address to the Geological Society 'has it ever been denied that this period may be enough for the purposes of geology' (Huxley, 1869, p. xlviii). In doing so, Huxley laid the seeds of discontent between the views of physicists in one corner and the geologists in the other corner. However, he did imply that the for the biologists Thomson's timescale didn't necessarily cause them problems, as they, the biologists, simply used the geologists timescale in the absence of having one for themselves.

In 1895, Thomson penned a short paper for *Nature* in which he discussed a recently published paper by Clarence King (1842–1901), who had been the first Director of the United States Geological Survey between 1879 and 1881. King's paper re-examined Thomson's work and using a new figure of 1,950°C for the fusion temperature of rocks, a figure which he had obtained from the physicist Carl Barus (1856–1935), arrived at twenty-four million years as the age of the Earth. This limit was accepted by Thomson who acknowledged that Barus' data would have rendered his 100-million-year limit too large, and that while the method would point to 10 million years, the effects of pressure on the geological processes would push the age determination to that of King's. In a short letter to *Nature* in 1897, Thomson settled on a range of 20–40 million years old but said that the 24-million-year estimate reached by King in 1893 was probably correct.

The cleric and geophysicist Osmund Fisher (1817–1913) stated in a paper published in the *Geological Magazine* in 1895 that 'no reliable estimate of the age of the world, based on considerations of the present temperature gradient at the surface, has hitherto been made' (Fisher, 1895, p. 246). James Geikie (1839–1915) Professor of Geology at Edinburgh wrote in the February 1900 issue of the *Scottish Geographical Magazine*, 'there are certain other considerations which increase one's doubts as to the adequacy of Lord Kelvin's theory' (Geikie, 1900, p. 63). Thomas Chrowder Chamberlin (1843–1928), President of the University of Wisconsin, was probably Thomson's most outspoken critic, at least in America Chamberlin, 1899). He believed that the Earth had formed thanks to the accretion of cold material and that it had never been never fully molten. Even the biologists were concerned that Thomson's age limits were too short for biological evolution. Edward Bagnall Poulton (1856–1943), Hope Professor of Zoology at Cambridge, weighed-in at the annual British Association meeting in Liverpool in 1896. As President of the Biological Section, he could create quite a stir, and in his address, he attacked the findings both of Thomson and those of some geologists (Poulton, 1897).

The main and most serious objection to some of Thomson's conclusions came from another Ulster Protestant, John Perry (1850–1920) (see Shipley, 2001; England *et al.*, 2007) who coincidently spent 1874 working as a research assistant in Thomson's laboratory in Glasgow before moving to Japan. Following a stint as a Professor at the Imperial College of Engineering in Tokyo, Perry returned to Britain. He took a position at the Finsbury Technical College in London and in 1896 moved to the larger Royal College of Science in the city.

In 1894, Robert Arthur Talbot Gascoyne-Cecil, 3rd Marquess of Salisbury (1830–1903) was President of the British Association for the Advancement of Science meeting held that year in Oxford. Although not a scientist, he was a Fellow of the Royal Society, and he was influential and people listened to what he had to say. He had recently twice served as Prime Minister, when his Conservative Party held power alternately with William Gladstone's Liberal Party, and he would hold this office again for a further eight years between 1895 and 1902. In his address, Salisbury attacked the basis of Darwin's theory of natural selection saying that there was not enough time for natural selection to have taken place. To back his assertion Salisbury depended on the age estimates and reputation of Lord Kelvin.

Having read the printed paper, Perry wrote to Kelvin three times and outlined a number of objections to his work and gave some suggestions as to how the calculations could be improved, but he received no response. Not content to be brushed off, Perry sought and received support for his objections from other scientists, and he then felt forced to commit himself to print. The following year his objections were outlined in two papers in *Nature*: in brief he said that convection, and not just conduction, played a role in heat loss from the Earth and that this invalidated Thomson's findings. Perry's objections were immediately countered by Kelvin in a note in the same journal, and this was followed by a third missive from the younger man. As Brian Shipley (2001, p. 94) has observed, Perry showed that:

> the faster heat was conducted outwards from the Earth's core, the longer it would take to obtain the present observed temperature gradient at the surface.

This was because Kelvin had based his calculations on the conductive properties of surface crustal rocks and he didn't include the denser sub-crustal rocks known to exist. Longer was counter to Kelvin's conclusion.

However, there was some support at this time, and it came from the American, George Ferdinand Becker, who in *Science* in February 1908 re-examined Thomson's methods and concluded:

> Notwithstanding the inadequacy of the data, I can not but believe that the 60-million year earth here discussed is a fair approximation to the truth and that with better data this age will not be changed by more than perhaps 5 million years.

Becker pointed out that this age was broadly concurrent with those derived from the sedimentation accumulation and oceanic sodium methods. Even Mark Twain (1835–1910), who wrote a short essay on geological matters, said of Thomson that 'I think we must yield to him and accept his view' (Quoted in Burchfield, 1990, p. ix).

Thomson's initial reasoning behind taking up research on the age of the Earth was, as we have seen, his difficulty with the Darwinian timescale, and Darwin himself acknowledged that it caused him trouble. In terms of his influence on the biologists who adhered to the longer time frame required by natural selection, it would appear to have been slight if we believe Archibald Geikie (1835–1924), Director of the Geological Survey, and later, like Thomson, President of the Royal Society. In fact, many biologists didn't require the longer timescale suggested by Darwin as many didn't accept natural selection. They preferred the Lamarckian means of biological change which required little time. Natural selection and thus the need for greater time were generally accepted by most biologists by the 1930s. In his Presidential address to the geological community at the British Association for the Advancement of Science meeting held in Dover in 1899, Geikie first acknowledged the debt geologists owed Thomson:

> Geologists have been led by his criticisms to revise their chronology. They gratefully acknowledge that to him they owe the introduction of important new lines of investigation, which link the solutions of the problems of geology with those of physics
>
> (Geikie, 1900, p. 6)

but he then delivered a sharp reprimand to the ennobled physicist and a reminder of the leanings of palaeontologists towards a chronology longer than that proposed by him:

> It is difficult satisfactorily to carry on a discussion in which your opponent entirely ignores your arguments, while you have given the fullest attention to his... [Geologists] have been willing to accept Lord Kelvin's original estimate of 100 millions of years as the period within which the history of life upon the planet must be comprised... yet there is undoubtedly a prevalent misgiving, whether in thus seeking to reconcile their requirements with the demands of the physicist they are not tying themselves down within the limits of time which on any theory of evolution would have been insufficient for the development of the animal and vegetable kingdoms
>
> (Geikie, 1900, p. 8)

Although critical of Thomson's work in the geological sphere, Geikie was rather friendly with him on a personal level and certainly respected his genius. In April 1899, he shared a railway compartment with Thomson and his wife on a journey to Genoa and the following day they explored the hills and scenery together. Geikie attended Thomson's funeral and in his autobiography remarked:

> ... we laid to rest all that was mortal of one who was not more reverenced for his towering genius and his manifold achievements in science, than he was beloved for the rare modesty, simplicity, and goodness of his character.
> (Geikie, 1924, p. 350)

Thomson's chronological conclusions and the demise of his timescale

Thomson's final contribution to the chronological debate was a paper entitled 'The age of the earth as an abode fitted for life'. This was published in several journals in England and America in 1898 and 1899 and is basically a summation of his ideas. He remarked that his determinations were based on science (i.e. physics) and haughtily concluded that if his methods proved incorrect then geologists and biologists were essentially free to plough their own furrows:

> If science did not allow us to give any estimate whatever as to whether 10,000,000 or 10,000,000,000 years is the age of the earth, then I think that Professor Huxley would have been perfectly right in saying that geologists should not trouble themselves about it, and biologists should go on in their own way, not inquiring into things beyond the power of human understanding and scientific investigation.
> (Thomson, 1899, p. 666)

He held on to his view that the Earth had consolidated no less than 20 million years ago and no more than 40 million years previously (but favoured the precise date of 24 million years suggested by Clarence King), and that the Sun was anything between 20 and 60 million years old (he favoured an age trending towards the lower time limit). One can see that Thomson's coeval age of both Sun and Earth coincided in and around 20–30 million years.

Thanks to his continual tinkering with his age determinations, Thomson more than likely reduced the credibility of his work, and his research in the area of geochronology was overtaken by the emergence of the study of radioactivity in geology. He failed to fully acknowledge this new science and see the full implications that it heralded, although interestingly he had in 1862 anticipated an internal heat source in the Sun:

> As for the future, we may say, with equal certainty, that inhabitants of the earth cannot continue to enjoy the light and heat essential to their life,

> for many million years longer, unless sources now unknown to us are prepared in the great storehouse of creation.
>
> (Thomson, 1889, p. 368)

Shortly before Thomson, by then Baron Kelvin of Largs, died, Ernest Rutherford (1871–1937), who was Professor of Physics at McGill University, Montreal, was waiting to give a lecture on his work to an assembled audience at the Royal Institution in London in 1904. Looking out from the lectern Rutherford was horrified to see the 79-year-old sitting in the front row, and he realised that the final section of his lecture would discredit a great deal of the older man's geological research on the age of the Earth. He had no option but to begin and was highly relieved shortly afterwards to see Thomson begin to slumber. Throughout the lecture, Thomson remained lost to the world but awoke at the critical point when Rutherford was about to deal with geochronology. He recalled later (see Eve, 1939, p. 107; Burchfield, 1990, p. 164):

> I saw the old bird sit up, open an eye and cock a baleful glance at me! Then a sudden inspiration came to me, and I said Lord Kelvin had limited the age of the earth, provided no new source of heat was discovered. That prophetic utterance refers to what we are now considering tonight, Radium! Behold! the old boy beamed upon me!

Initially, while Thomson recognised that the radioactive material Radium produced heat, he refused to believe that it produced it itself, and rather argued that the Radium must have gained the heat that it subsequently emitted from an external source. However, soon afterwards, he privately accepted that the discovery of Radium had made some of his conclusions regarding secular cooling of the Earth difficult, though he never made this view publicly known. Perhaps, it would have been best if he had slept through Rutherford's lecture, but there is little doubt that he saw the dawn of radioactivity herald the demise of his own geochronology.

In public, Thomson remained defiant to the end of his life. None of the criticisms of his methods thrown at him by geologists or the biologists made him moderate or alter his views substantially and he continued to hold that the Earth was rather young.

1900s

16

HISTORY OF ICHNOLOGY: JOHN JOLY (1857–1933) ON *OLDHAMIA*: POETIC AND SCIENTIFIC OBSERVATIONS

Ichnos 18 (2011), 209–211.

Introduction

In the 1880s, the Irish geologist and geophysicist John Joly (1857–1933) (Figure 16.1) wrote a sonnet titled 'Oldhamia antiqua'. Here, this paper draws from his poem some geological and geochronological meaning and briefly discusses Joly's only scientific foray into ichnology.

Joly was born in King's County (now Co. Offaly) in the Irish midlands in 1857, the youngest son of an Anglican cleric. He was educated at Trinity College, Dublin, where he graduated with a degree in engineering before taking up a series of assistantships in engineering and physics in the university. He was appointed as Professor of Geology and Mineralogy in 1897. He published on a wide diversity of topics, including geology, mineralogy, geophysics, tectonics, optics, chemistry, photography, and mechanics, and he notably invented a number of important pieces of equipment such as the photometer, a mercury-glycerine barometer and a steam calorimeter for measuring the specific heat of minerals (Wyse Jackson, 1992a). John Joly can rightly be regarded as Ireland's foremost scientist of the early twentieth century. Further comprehensive details of his life and work are given in Dixon (1934), Nudds (1986), Pemberton (2010), and Wyse Jackson (1992a, 2007c).

Inspiration under canvas

In the early to middle 1880s, Joly and a circle of friends took up camping, or 'tenting' as they put it, as a means of recreation. They ventured into the hills and mountains of south county Dublin and north Wicklow, which were within easy reach of Dublin city where they worked (H.H. Dixon, unpublished chronology of Joly's life, Geology Department archives, Trinity College, Dublin). Being observant and having a wide-ranging curiosity, Joly was enthralled by the geological

 DOI: 10.4324/9781003475552-20

Figure 16.1 John Joly in his early twenties (Author's collection).

landscape, and he began to collect interesting geological specimens, including fossil and petrological material. In particular, he developed an interest in acquiring minerals and over his lifetime assembled a reasonably diverse collection of specimens (Wyse Jackson, 1992a). During his sojourns into the depths of Dublin and Wicklow, he was also inspired to pen poetry, and his sonnet 'Oldhamia antiqua' was written on 28 August 1886, while under canvas (Nudds, 1983).

This period in the early to mid-1880s was significant, as it marked Joly's first excursions into geological research. This work was to have major consequences for his future career. Aside from his two short papers on the trace fossil *Oldhamia* (Joly, 1886a, c), he also reported on the minerals Beryl and Iolite (Joly, 1885) and Harmotome from counties Dublin and Wicklow (Joly, 1886b), and he produced a longer paper on the specific heats of minerals (Joly, 1886d). In 1897, William Johnston Sollas, author of the ichnogenus *Pucksia* described from the Cambrian of Howth, Co. Dublin (Sollas, 1895a), resigned as Professor of Geology and Mineralogy at Trinity College, Dublin. Although Joly had spent all his career to that point as an Assistant to the Professor of Engineering and then Physics, he argued in a testimonial to the Board of the College that he had the necessary geological expertise that made him an ideal candidate for the geological position. His application was successful, and he went on to make major geological advances in geochronology, radioactivity, and tectonics (Wyse Jackson, 2001a [Chapter 17 herein], 2002a, 2006a), although he did not return to ichnology.

Oldhamia antiqua: Joly's poetic observations

Is nothing left? Have all things passed thee by?
The stars are not thy stars! The aged hills
Are changed and bowed beneath repeated ills
Of ice and snow, of river and of sky.
The sea that raiseth now in agony
Is not thy sea. The stormy voice that fills
This gloom with man's remotest sorrow shrills
The memory of the futurity!
We – promise of the ages! – Lift thine eyes,
And gazing on these tendrils intertwined
For Aeons in the shadows, recognize
In Hope and Joy, in heaven-seeking Mind,
In Faith, in Love, in Reason's potent spell
The visitants that bid a world farewell!

This poem was first published in Dixon (1934) and reprinted by Nudds (1983), Wyse Jackson (2001a), and Pemberton (2010). None of these authors attempted a detailed analysis of its meaning. In this paper, the sonnet's meaning and the various influences on its author are considered fully for the first time. The main interpretation of Joly's thoughts provided in this sonnet is that he is contemplating time, as well as geological processes, and the possibility of an overall deity. Joly recognises that geological history is infinitely long. At the time he penned these lines, his concept of geological time, as was the case with the majority of his scientific contemporaries, was based by the then generally prevailing views of William Thomson, who estimated that the Earth was formed 40 million years ago. The lines 'The stars are not thy stars!' and 'The sea that raiseth now in agony / Is not thy sea' suggest the passage of a long period between when *Oldhamia* was living and when Joly found the ichnofossil. Joly fully understood the relative position of the Cambrian in the geological succession, and it was only later in 1900 that he returned to the longevity of the oceans, when he calculated that the oceanic waters had taken 100 million years to attain their present saltiness and that this meant the Earth was that old. Later, with Ernest Rutherford, he determined the Leinster Granite that crops out close to Bray Head in County Wicklow just south of Dublin to be 400 million years old (Wyse Jackson, 2001a).

Further, one deduces that in this piece Joly also addresses denudation and to a lesser extent oceanography, given that he mentions the sea. Standing on the prominent crest of Bray Head that extends into the Irish Sea and looking south-west towards the rolling Wicklow Mountains, Joly understands that the landscape he is familiar with is constantly undergoing change: 'The aged hills / Are changed and bowed beneath repeated ills.' This change is affected, Joly notes, by various factors: 'Of ice and snow, of river and of sky.' These processes sculpted the Bray rocks and revealed the ancient fossil *Oldhamia* to him: 'And gazing on these tendrils intertwined / For Aeons in the shadows.'

In the ninth line of the sonnet, Joly seems to be saying that man is the ultimate evolutionary design—'We-the promise of the ages'—and that man has a responsibility to acknowledge these earlier inhabitants of our globe. He argues that we should draw comfort from these early life forms as part of an overall scheme of which man is part and who has a 'heaven-seeking Mind', thus a religious element to our makeup.

Oldhamia: Joly's scientific observations

During his Christmas vacation in 1884, Joly spent some time walking in parts of County Wicklow (Dixon, unpublished chronology of Joly's life). There, on some early morning rambles, he observed particular markings in the muddy paths which reminded him of *Oldhamia radiata* that he had collected from Bray Head. He inferred that these markings were left when the frost that had made them had melted. Naturally enough, he wondered if the mechanism that had produced the modernmarkings could be responsible for the ancient Cambrian traces. He explained the formation of the pattern in the modern mud noting that the larger and more prominent grains in the mud would become centres of crystallisation of the interstitial water and that the grains would become aligned parallel to the propagating ice spicules (Joly, 1886a, p. 157).

Joly then tried to replicate these conditions in the laboratory by subjecting layers of mud to frost, and he successfully generated patterns that resembled *Oldhamia radiata*. However, he failed completely in generating patterns that resembled *Oldhamia antiqua*. He attributed this failure to problems with the texture of the muds in his experiment and thought that he might be successful if he were to use mud derived from crushed Cambrian rock from Bray. Unfortunately, due to a lack of time, he never continued with this research.

He concluded that it was possible that *Oldhamia* was inorganic in origin and had been produced by a physical process when muds were exposed at low tide and subjected to freezing. Of course, we now know that the Cambrian shales and greywackes were deposited in deep water and that subaerial exposure would not have occurred at any time during the formation of *Oldhamia*.

Subsequent to this work, Joly penned the sonnet discussed above. The phrase used in line ten—'these tendrils'—strongly suggests that he rejected his non-organic hypothesis. Late in 1886, he read his second paper of that year on *Oldhamia* to the members of the Royal Dublin Society, and this questions the genetic similarities of two ichnospecies (Joly, 1886d). Having examined twelve *Oldhamia*-bearing slabs in his possession, Joly noticed that where one species was preserved as a positive the other was always preserved as a negative (Figures 16.2). He then proceeded into the field where he collected more specimens and found that '*O. radiata* appeared invariable as a depression, the

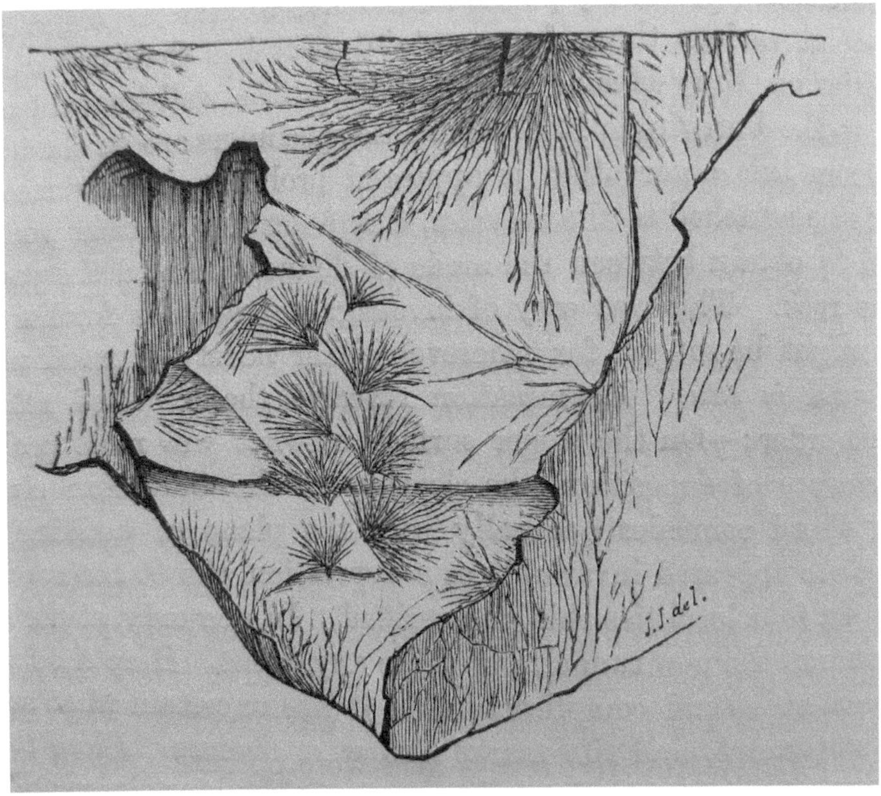

Figure 16.2 Woodcut illustration drawn by John Joly of *Oldhamia antiqua* (bottom left) and *Oldhamia radiata* (top) (From Joly, 1886c; Geological Museum, Trinity College Dublin).

O. antiqua in relief' (Joly, 1886c, p. 445). This finding perplexed him, and he concluded that these differences could be attributed to the relationship between the bedding plane on which they are preserved and the mechanism by which they were produced. He did not clarify what the relationship was between these two parameters. However, he accepted not only that these findings rendered his frost-formation hypothesis void, but also that an organic origin for the two forms could not be discounted.

Conclusions

John Joly's ichnological legacy can be found in the pages of the *Treatise of Invertebrate Paleontology*, where the specimen of *Oldhamia antiqua* that he collected from the Cambrian of the Bray Group is illustrated (Häntzschel, 1962, figs. 126.6, 1975, figs. 53, 3b). This and the specimen illustrated by him in his 1886 paper

(Joly, 1886c) testify to Joly's attempts from the perspective of an engineer to explain the formation of this trace fossil in a logical manner by analogy with the growth and crystallisation of frost. He later changed this view and thought it may be organic in origin.

While *Oldhamia* remains somewhat enigmatic and Joly's interpretations of its formation do not stand up today to close scrutiny, it is not difficult to understand why this trace fossil inspired him to poetical writings.

17

JOHN JOLY (1857–1933) AND HIS DETERMINATIONS OF THE AGE OF THE EARTH

In C.L.E. Lewis and S.J. Knell (eds). 2001.
The Age of the Earth: from 4004 BC to AD 2002.
Geological Society Special Publications, 190
(2001), 107–119.

Introduction

John Joly was born on 1 November 1857 in Hollywood House (the Rectory), Bracknagh, County Offaly, the third and youngest son of the Reverend John Plunket Joly (1826–1858) and Julia Anna Maria Georgina née Comtesse de Lusi. The Joly family originated from France but came to Ireland from Belgium in the 1760s. Joly's great-grandfather served as butler to the Duke of Leinster who gave the living of Clonsast parish to the family in the early 1800s (Dixon, 1934; Nudds, 1986).

After his father's sudden death at a young age, the Joly family moved to Dublin where John Joly received his secondary education at the celebrated Rathmines School in which he was enrolled from 1872 to 1875. Although he did not excel in the classroom, he was popular, nevertheless, and became known as 'The Professor' on account of his tinkering with chemical apparatus and other gadgets. In 1875, after a bout of poor health Joly spent some time in the south of France, before returning to Dublin in 1876. He then entered Trinity College, Dublin, where he remained for the rest of his life. He followed courses in classics and modern literature but later concentrated on engineering. In 1882, he sat for the degree of Bachelor of Engineering and gained first place and certificates in several subjects including engineering, experimental physics, mineralogy, geology, and chemistry. In adulthood, Joly was a distinctive and unforgettable man. Tall, with hair swept off his forehead, a bushy moustache, and pince-nez perched on his nose, he spoke with what was considered to be a foreign accent, but in reality, the rolled r's were simply used to conceal a slight lisp (Dixon, 1941). He was a keen traveller and made many trips, especially to continental Europe and the Alps, with his life-long friend Henry Horatio Dixon (1869–1953), one-time Professor of Botany at Trinity

DOI: 10.4324/9781003475552-21

College Dublin. Joly loved the sea and was a notable yachtsman who made frequent voyages along the western seaboard of the British Isles. He also served as a Commissioner of Irish Lights and carried out some work for the Admiralty on signalling and safety at sea (Nudds, 1986, 1988).

Academic career

Although Joly considered a career abroad, he acceded to his widowed mother's wishes and remained in Ireland, at Trinity College, Dublin. There he was employed to carry out some teaching and to assist in the research of the professor of civil engineering. During this time (1882–1891), he was able to engage in a great deal of research, although this was largely in aspects of physics and mineralogy, rather than engineering. As with most experimental research of this period, apparatus did not exist, or was not available, and so had to be invented and built by the researcher himself. Among the first pieces invented by Joly were a new photometer and a hydrostatic balance. It was during this time that he developed an interest in mineralogy and began to accumulate a large collection of very fine Irish, continental and American mineral specimens (Wyse Jackson, 1992a). These he bequeathed together with the cabinet in which he housed them to Trinity College Dublin where they are now in the Geological Museum (https://trinitygeologicalmuseum.com/collections/joly-collection/). He invented the steam calorimeter for measuring the specific heat of minerals. At the same time, he worked on the density of gases and the detection of small pressures, and the steam calorimeter later played an important role in the kinetic theory of gases (Somerfield, 1985).

In 1891, Joly was appointed assistant to George Francis Fitzgerald, Professor of natural and experimental philosophy, and the following year, he was elected a Fellow of the Royal Society, an indication of how important his contemporaries considered his work to be. Surprisingly, he was not admitted as a member of the Royal Irish Academy for another five years. Over the ensuing years, he addressed a tremendous range of scientific subjects. In 1894, together with Dixon, he published an important paper that explained for the first time the mechanism of the ascent of sap in trees and plants (Dixon and Joly, 1895, and in 1895, Joly exhibited colour slides to the Royal Society. His method, the 'Joly process of colour photography', was the first successful process of producing colour photographic images from a single plate (Joly, 1896). However, his assertion that he had invented the technique was challenged by an American. Legal battles ensued but although Joly's priority was finally established, the process was soon superseded by other methods. Nevertheless, the 'Joly process' is important, as it is essentially the method used in colour photography today.

In 1897, Joly succeeded William Johnston Sollas in the chair of geology and mineralogy at Trinity College, a position he retained until his death in 1933. Unlike his predecessors, Joly was essentially a physicist, not a geologist, and consequently, he published little on the geology of Ireland. He carried out much research on minerals, but it was his work on radioactivity and radium that was his most important.

Joly pioneered the medical benefits of radioactivity. This led to his establishing, with a Dublin doctor, Water Clegg Stevenson (1876–1931), of the Irish Radium Institute in 1914, which exploited the medical advantages of the radioactive element in treating cancer patients (Joly, 1931). Their 'Dublin method' was the first to utilise radium emanation (radon) enclosed in hollow needles in the treatment of tumours (Murnaghan, 1985). Joly was very proud of this work.

But Joly also worked on determining the age of the Earth, and it is for his contributions to this debate that he is now best remembered.

Joly's sodium method

Joly's contribution to the question of the age of the Earth was one of a number made by Irish-born or Irish-based academics (O'Donnell, 1984; Wyse Jackson, 1992a) (Table 1). The first was that of Archbishop James Ussher, who in the seventeenth century arrived at a date of 4,004 BC from a reading of biblical events (Reese *et al.*, 1981; Brice, 1982; Fuller, 2001). Others included Belfast-born William Thomson (later Lord Kelvin) who argued, based on the cooling rate of an initially molten planet, that the Earth was between 20 and 400 million years old (Thomson, 1864). However, his last paper on the subject, published in 1899, firmly placed the limits between 20 and 40 million years (Thomson, 1899), a value that antagonised many geologists (Dalrymple, 2001; Shipley, 2001). Samuel Haughton and Sollas, two of Joly's predecessors at Trinity College, Dublin, also entered the geochronological debate, and their contributions are discussed below.

Joly's first documented thoughts on the antiquity of the Earth were in written verses penned on 28 August 1886 in the Wicklow Mountains south of Dublin (Nudds, 1983; Pemberton, 2010; Wyse Jackson, 2011 [Chapter 16 herein]). One sonnet considered the age of the enigmatic trace fossil *Oldhamia antiqua* Forbes from the Cambrian slates of Bray Head, which Joly suggested was a witness to the long, slow changes that had affected the Earth.

Joly's first scientific foray into the matter of geochronology came thirteen years later with the publication of his first and probably most celebrated, if somewhat controversial, paper on the subject (Joly, 1899). Simply put, Joly examined the rate of sodium input into the oceans and by simple mathematics arrived at an estimate for the age of the Earth. The idea came to him whilst sailing off the east coast of Ireland with Henry Dixon in 1897, collecting coccoliths and plankton from the Irish Sea (Dixon and Joly, 1897, 1898) and observing the feeding behaviour of seabirds (Joly, 1898).

At that time, Joly was unaware, as were his contemporaries, of the pioneering work of the English astronomer Edmond Halley (Cook, 1998). In 1715, Halley had proposed to the Royal Society that salt concentrations in lakes that had no discharge rivers, should be measured every hundred years, as he considered that from the incremental increase of the salt, the age of the lake could eventually be deduced. Once enough data had been collected over time, inferences about the age of the ocean, and therefore of the Earth, could be drawn from the results.

While Halley accepted that mankind had dwelt upon the Earth for about 6,000 years, as stated in the Scriptures, he also considered that it was 'no where revealed in Scripture how long the Earth had existed before this last Creation' (Halley, 1715, p. 296). He thus regretted that the ancient Greek and Latin authors had not recorded the saltiness of the sea 2,000 years ago, since 'the World may be found much older than many have hitherto imagined' (Halley, 1715, p. 299). He therefore recommended to the Society that experiments be started 'for the benefit of future Ages'. But the Society does not seem to have heeded his advice and Halley's idea was only rediscovered in 1910 (Becker, 1910a).

Although aware of Thomas Mellard Reade's important book of 1879, Joly was unaware of his valuable paper published in the *Proceedings of the Liverpool Geological Society* in 1877, which examined the volume of sulphates, carbonates, and chlorides in the oceans, and their rate of accumulation. Using this information Reade calculated the time taken for the oceans to reach their present concentration of these substances: 25 million years, nearly half a million years, and 200 million years, respectively, for sulphates, carbonates, and chlorides. These figures gave an estimate of the minimum age of the Earth (Mellard Reade, 1877, p. 229). Although Joly later acknowledged these pioneering publications of Halley and Mellard Reade he noted, without explanation, that their schemes, unlike his, would not have produced reliable results (Joly, 1915).

Brilliant in its simplicity, Joly's paper of 1899 fired the imagination of both scientific and general audiences, and for perhaps a decade his 'sodium method' held sway amongst geochronologists. It relied upon a number of assumptions, and on data regarding ocean and river characteristics published by John Murray in the 1880s. The fundamental tenet was the uniformitarian stance that the rates of denudation of sodium-bearing rocks, and the discharge of the rivers into the oceans, had remained uniformly constant over geological time. The age of the Earth was thus derived by the simple formula:

Earth's age = Volume of sodium in the ocean / Rate of annual sodium input

which yielded an age of 90–100 million years.

Joly read his paper to the Royal Dublin Society at a meeting in Leinster House on 17 May 1899, and it was published several months later in September in the *Scientific Transactions*, the premier journal of the Society. Reactions to his ideas began to appear in the scientific press within six months. Review articles were published in several journals, including the *American Journal of Science* (Anon., 1899) and *Geological Magazine* (Fisher, 1900), but the Reverend Osmund Fisher's review was by far the most testing, because, he argued, the processes invoked by Joly were not uniform throughout geological time. Additionally, Fisher suggested that Joly's figures for the volume of sodium delivered into the oceans by rivers might be at fault, and that Joly did not take into account the effect of 'fossil sea water' which, Fisher noted, was trapped in sediments and elsewhere. Fisher also suggested that Joly's estimate, that 10% of the sodium chloride came from rainwater, was too high. Finally, he made a little swipe at Joly's written style which he contended was rather convoluted and not very clear. He has a point.

These minor criticisms aside, Fisher recognised that Joly's essay had 'opened up an entirely new line for the investigation of geological time' (Fisher, 1900, p. 132).

William Ackroyd, Public Analyst for Halifax, was of the opinion that a great deal of the oceanic salt was transported back onto land; he put this figure at 99%, and on the basis of this recalculated the age of the Earth and derived an age of 8,000 million years (Ackroyd, 1901a). This was the first offering in a public debate on Joly's theory played out in the pages of *Chemical News* and *Geological Magazine* (Ackroyd, 1901a, b; Joly, 1901b, c). Ackroyd argued that it was important to know the ratio in river water of transported (or recycled) sea salt and that derived from rocks through solvent denudation. He included data that demonstrated that only 0.02% of the chlorides in the water of Malham Tarn near Craven were derived from the surrounding limestones. From this, one must assume he considered that most were derived from atmospheric water. In his papers in *Geological Magazine* (Ackroyd, 1901b), he criticised Joly for not appreciating the significance of this work or for ignoring it altogether.

At the British Association for the Advancement of Science meeting held in 1900 at Bradford, Joly's findings must have created quite a stir as his report was ordered by the general committee to be published *in extenso* (Joly, 1900c). William Sollas, in his presidential address to Section C, Geology, sided with Joly's finding, stating that 'there is no serious flaw in the method, and Professor Joly's treatment of the subject is admirable in every way' (Sollas, 1900). Sollas, however, did question the reliability of the data concerning the river discharge of sodium (Sollas, 1900). Later papers by Sollas (1909), Becker (1910b), and Clarke (1910) laid minor criticisms at Joly's door. Sollas suggested that some modification could be made for the fact that volcanic activity was at certain times in the past more pronounced that at present times, and that this would have had some effect on the supply of sodium to the oceans. In addition, he argued that modern ocean water, at normal temperature, had only a slight corrosive effect on salt contained in rocks. At higher temperatures of between 180°C and 370°C far higher volumes of sodium would have been dissolved. From this, we surmise that Sollas believed that the early oceans were considerably hotter than those of today. Sollas recalculated the annual discharge of the rivers from which he derived a date of 78 million years for the age of Earth but suggested that it lay within the range of 80–150 million years. Clarke (1910) examined the rate of removal of sodium from the landmass and arrived at a figure of 80 million years, while Becker (1910c) suggested that Joly's figure of 10% for the contribution of sodium recycled from the atmosphere was closer to 6%. Subsequently, a considerable number of papers also discussed Joly's sodium method (for example, Rudzki, 1901), which were either in broad agreement with his ideas (e.g. McNairn, 1919) or raised a number of objections (see for example Shelton, 1910, 1911; Holmes, 1913). Shelton said that while Joly's scheme was instructive, some of the underlying foundations on which it was based needed careful consideration. Joly, Shelton argued, did not make adequate allowance for salt recycled via the atmosphere, nor for fossil salt contained in sedimentary rocks. Shelton also noted that Joly's age determination was based on analyses of sodium content, which given the accuracy of instrumentation available could not be done with any great accuracy. In particular, Arthur Holmes reasoned that the rocks would have had to lose more sodium into the oceans than they had ever contained, for Joly's figures to add up.

Joly's responses to these criticisms (Joly, 1900b, 1901b, c) strongly reinforced his uniformitarian principle; however, he did accept that his estimate of the role of rainwater in providing sodium chloride might have been underestimated and required further experimental work. With respect to fossil seawater, Joly stated that it could only have contributed 0.9% of oceanic sodium chloride and, as such, was negligible. His response to Sollas was that 'there is much reason to believe that the nineteen rivers… afford an approximation as to what the world's rivers yield' (Joly, 1900c). Indeed, he stated in 1911 that the findings of Sollas, Becker and Clarke, together with his own, gave concurrent results of *c.* 100 million years, and proudly anticipated that this determination would not be 'seriously challenged in the future' (Joly, 1911a).

Joly went further in defence of his ideas, in that he devised various experiments which he hoped would generate acceptance of some of his theoretical assumptions made in 1899. One of these was to invent a fractionating rain gauge (Joly, 1900a), which he hoped would allow him to collect rainwater over incremental time periods. Subsequent analysis of the amount of dissolved sodium chloride in this rainwater would enable him to quantify the volume of sodium in the oceans from this pluvial source. While demonstrating how his rain gauge would operate, Joly fails to record whether it was ever put to use or was effective!

Joly also examined the rate of solution of various igneous materials in fresh and salt water (Joly, 1900d) which showed that of the four tested (basalt, orthoclase, obsidian and hornblende) the basalt from the Giant's Causeway in County Antrim dissolved more readily than the others, and that salt water was a more effective solvent than fresh water. Not surprisingly, the obsidian proved the most resistant to solution. Joly noted that his results for the rates of denudation were far lower than those demonstrated by field study and he argued that additional factors such as organic acids, wetting and drying, and other erosive processes were more important than solution of rocks by water. Nevertheless, he made an allowance for the solvent action of the ocean, by reducing his age estimate by a few million years to 96 million (Joly, 1911a).

After the initial peak of interest that closely followed on from his 1899 paper, many of Joly's subsequent papers on the subject were simply reports of lectures, or reiterations of the original theory. In 1915 with the publication of *Birth-time of the World* (Joly, 1915), there followed renewed interest in the sodium method—but it did not last. The theory was finally consigned to the scientific scrap heap by several geologists (Haeker, 1914; Barrell, 1917; Gregory, 1921; Chamberlin, 1922) who, in the damning words of Arthur Holmes, 'rejected it as worthless' (Holmes, 1926, p. 1056). By the mid-1920s, the scientific community was focused on the new theories based around radioactivity (Lewis, 2001). Paradoxically, Joly also carried out much useful research in this developing area, but he himself could never consign his sodium method to the waste basket, although by 1930, he had accepted some major modifications suggested by Alfred Church Lane in 1929, which pointed at a figure of 300 million years for the method (Joly, 1930).

Unusually, and perhaps uniquely for publications of the Royal Dublin Society, a second impression of the original paper had to be produced in November 1899 as all the stocks had been distributed and demand continued. This allowed Joly to rectify a number of small errors that had appeared in the appendix of the first impression. The paper was also printed in North America in its entirety in the *Annual Report of the Smithsonian Institution* for 1899, and so Joly's ideas and methodologies were rapidly transmitted throughout the scientific community on either side of the Atlantic (Joly, 1901a). It was an important contribution to the growing body of scientific opinion that refuted the low estimate of the age of the Earth of 20–40 million years propounded by Kelvin (see, for example, Shipley, 2001), which was eventually dispelled with the advent of radioactive dating methods.

It is interesting to note that it is now thought that on the basis of the volume of chloride contained in brines found in deep-seated groundwater, the salinity of the earliest oceans is considered to be 1.5–2 times saltier than that of today (Knauth, 1998a, b). Thus, the oceans are not becoming progressively saltier through the release of sodium and chloride during denudation, as was Joly's contention. Quite the contrary, Knauth (1998a) argued that if all the present deposits of subsurface salt were returned to the oceans, they would be 30% saltier than at present.

Sediment accumulation

At the same time as the 'sodium method' was gaining acceptance, others continued to estimate the age of the Earth using sediment accumulation as their gauge. This method owed its origins to the work of John Phillips in 1860 (Phillips, 1860; Morrell, 2001) who determined the thickness of the global sedimentary pile and, using a figure for the rate of sedimentation, arrived at 96 million years for the age of the Earth. Many other calculations using this method followed (see Lewis, 2001), perhaps, the most celebrated of which were those of the Reverend Samuel Haughton, third professor of geology at Trinity College, Dublin. Haughton, who was a supporter of Kelvin's methods for estimating the age of the Earth and an opponent of Charles Darwin's theory of evolution, initially achieved an astonishing estimate of 2,298 million years, based on the same principles as those used by Kelvin (Haughton, 1865). In the context of sediment accumulation, however, Haughton is remembered for his principle that 'the proper relative measure of geological periods is the maximum thickness of the strata formed during these periods' (Haughton, 1878), which of course necessitated a global estimate of the maximum thickness of each sedimentary sequence, and the determination of the rate at which those sediments accumulated. Using this principle, he first published an age for the Earth of 1,526 million years (Haughton, 1871a), but uncomfortable with the vast timescales he was deriving, which were in stark contrast to those deduced by Kelvin, seven years later he attempted the calculation again and arrived at much the same answer. This time, however, he conceded:

> If we admit (which I am by no means willing to do) that the manufacture of strata in geological times proceeded at ten times this rate, or at the rate of one foot for every 861.6 years This gives for the whole duration of geological time a minimum of two hundred million years
>
> (Haughton, 1878, p. 268)

a value much more acceptable to the wider geological community. Much later Arthur Holmes argued that the uniformity of sedimentation rates assumed by Haughton was incorrect and that his principle would be better stated as 'the time elapsed since the end of any geological period is a function of the sum of the maximum thicknesses accumulated during all the subsequent periods' (Holmes, 1947, p. 119).

By the 1890s, the methodology had become the standard means of measuring the age of the Earth and a great many authors attempted it, including Charles Doolittle Walcott in the United States who derived a date of 35–80 million years based on measured sections in North American sedimentary basins (Walcott, 1893) and Mellard Reade in England who estimated that 'The time that has elapsed since the commencement of the Cambrian is therefore in round figures 95 millions of years' (Mellard Reade, 1893, p. 100). In 1895, Sollas, following Haughton's principle, calculated the Earth's age to be 17 million years (Sollas, 1895b), one of the lowest figures ever established. However, a later calculation, based on a total sediment thickness of 335,000 feet or 63 miles, and sedimentation rates of 3 and 4 inches per century resulted, respectively, in ages of 148 and 103 million years (Sollas, 1909). Sollas noted, though, that it was difficult to determine accurately the rate of sediment accumulation, which he acknowledged could be anything between 2 and 12 inches per century.

Joly was delighted since these dates largely concurred with his and thus confirmed the strength of his sodium method determination. In 1909, he examined Sollas' figures for himself and agreed with his results (Joly, 1911a). However, in 1914, in a lecture to the Royal Dublin Society, he argued that sediment mass, not thickness, was a more accurate measure, which widened Sollas' results to a minimum of 47 million years and a maximum of 188 million years. Then, on the basis that, he believed sedimentation rates were not uniform through geological time (Joly, 1915), he reduced the mean of these limits, 117 million years, to a figure of 87 million years which concurred well with his sodium method results. This marked a change in the uniformitarian stance that he had adopted in 1899 for the rate of sodium accumulation in the oceans.

By 1910, these methods were being supplanted by the age determinations generated by radioactive decay methods. Although in its infancy, the study of radioactivity was beginning to yield ages for the Earth that were considerably older than 100 million years.

Radioactivity and pleochroic halos

Joly was just one of many scientists drawn into the field of radioactivity following its discovery by Henri Becquerel in 1896 and the subsequent discovery of radium by Marie and Pierre Curie in 1898. By 1904, Ernest Rutherford had already suggested that the presence of radioactive minerals might provide a measure of the age of those minerals (Rutherford, 1905), and by 1905, Robert Strutt (later Lord Rayleigh) reported widespread presence of radioactive elements in rocks. Over the next five years, he went on to examine helium as a means of determining ages. In the United States, in 1907 Bertram Boltwood was the first researcher to examine the uranium-lead series (Badash, 1968; Dalrymple, 2001) and, using this method, in 1911 Arthur Holmes published a date of 1,640 million years for the Archaean of Ceylon (Holmes, 1911; Lewis, 2001). Immediately, Joly was on the back foot defending his geochronology from these new data. In his book *Radioactivity and Geology* (Joly, 1909c) and in an important paper entitled 'The Age of Earth', Joly (1911a) questioned whether the decay rate of uranium had been constant throughout geological history, as suggested by others. He said that this assumption was without strong basis and that the calculations of age limits were based on derived radioactive products, rather than radioactive parent elements. Turning to his own previous methods for support of his unease, Joly noted that the extremely slow rates of sediment accumulation implied by these vast ages were difficult to credit:

> If the recorded depths of sediments have taken 1400 million years to collect, the average rate has been no more than one foot in 4000 years! This seems incredible: and if we double the depth of maximum sedimentation it still remains incredible. But, if possible, still more incredible is the conclusion respecting solvent denudation to which radioactivity drives us. If the sodium in the ocean has taken 1400 million years to accumulate, the rivers are now bearing to the sea about 14 times the average percentage of the past It seems quite impossible to find any explanation of such an increase.
>
> (Joly, 1911a, p. 379)

Joly was not alone in his concerns the American geochronologist George Becker also voiced unease with radiometric dates, as did the American Committee on the Measurement of Geological Time by Atomic Disintegration, which reported that uranium as a whole decayed more rapidly in the past, and therefore the dates derived from it could be overestimated by 25% (Lane, 1925). Lane was in fact ahead of his time in recognising that uranium might have several isotopes (not confirmed for another four years, and U235 was not discovered until 1936) and thus 'as a whole' (averaging all isotopes) uranium appeared to have decayed more rapidly in the past. U235 decays six times as fast as U238 Holmes (1913), though, considered it highly unlikely that radioactive decay rates had varied through geological time: as unlikely as finding that the laws of physics and chemistry had changed over time!

Joly's first paper on radioactivity and geology was published in *Nature* in 1903 where he discussed the potential of using radium to date the age of the Earth (Joly, 1903). In a number of later papers published in 1908 and 1909, he calculated the volume of radioactive elements, including radium and thorium, in terrestrial and oceanic rocks of various ages (Joly, 1908b, 1909a), and from seawater (Joly, 1908a, 1909b). Much of this work received widespread release in his book *Radioactivity and Geology* (Joly, 1909c), and the outcome of this study was the formulation of his ideas pertaining to internal heat sources in the Earth, which had implications for a later geochronological method. It would be interesting to re-examine Joly's rocks from the Simplon and St Gothard tunnels through the Alps (Joly, 1907b, 1912), now housed in the Geological Museum in Trinity College, Dublin, to determine how accurate his results were. Do they match with modern calculations of the radioactive elements contained in these rocks? If not, what does this say about Joly's methodologies or the reliability of his equipment? Perhaps, as Léo Laporte has pointed out, any differences may result from the different half-lives then in use.

In 1907, Joly realised that small dark rings, or pleochroic halos as they came to be known, which he had observed in biotite in some granites, were the products of radioactive decay in zircons enclosed within the biotite crystals (Joly, 1907a). Previously, it had been suggested that these were due to the presence of organic pigments in the minerals, but Strutt had earlier demonstrated the radioactive properties of zircon, to which Joly attributed the halos. The size of the halo was related to the type of radioactive decay product and the range of the rays produced (Joly, 1911b) (Figure 17.1), while the intensity, he argued, was due to the duration of radioactive decay. He observed complex halos with distinctive inner and outer rings, or corona, in a greisen from Saxony in Germany (Joly, 1910), and soon afterwards, he and his research assistant, Arnold Lockhart Fletcher, attributed the development of the outer rings to the alpha decay of radium C (now known as the isotope bismuth^{214}) while the inner rings were produced by radium A (now known as the isotope polonium^{218}) (Joly and Fletcher, 1910). Joly was also able to distinguish halos produced by other radioactive sources including thorium and uranium, and the radii of the halos produced by the thorium and uranium decay series were given in the Huxley Lecture that Joly delivered in Birmingham in October 1912 (Joly, 1913).

In 1913, Joly and Ernest Rutherford developed a unique methodology to date a rock based on its pleochroic halos, which required knowledge of the mass of the nucleus of the halo and the number of alpha rays required to produce a certain intensity of halo. Using specimens from the Leinster Granite of Co. Carlow, Rutherford, working in Manchester, produced artificial halos in mica and measured the number of alpha rays required to produce them. Meanwhile, in Dublin, Joly measured the mass of the nuclei. Between them, they tabulated the ages of 30 halos which ranged from 20 to 470 million years. They concluded that the age of the Devonian was not less than 400 million years, which concurred with results found by Holmes two years earlier; however, a proviso read:

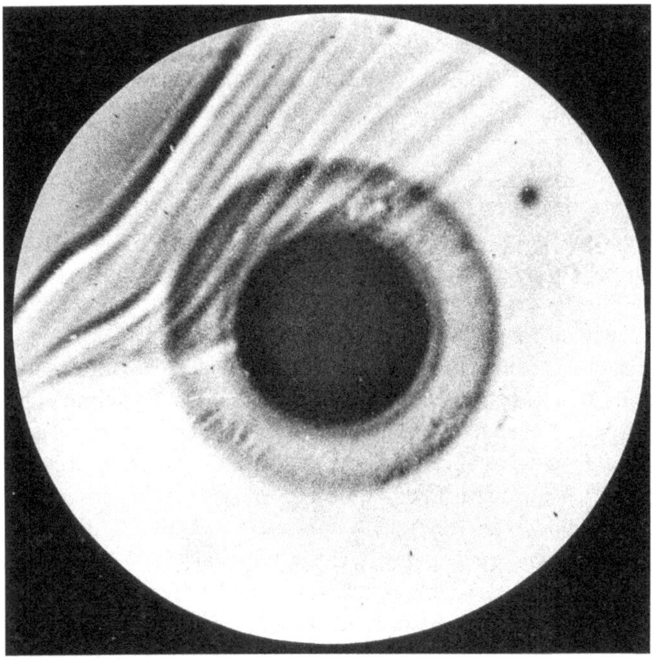

Figure 17.1 A single pleochroic halo developed in biotite in the Leinster granite from Garryellen, County Carlow, Ireland. The inner dark disc is due to radon (a gas derivative of radium); the succeeding inner ring is due to radium A alpha decay while the outer darker ring was produced by radium C alpha decay (From Joly, 1911b, plate III, fig. 4; Geological Museum, Trinity College Dublin).

> that if the higher values of geological time are so found to be reliable, the discrepancy with estimates of the age of the ocean, based on the now well-ascertained facts of solvent denudation, raises difficulties which at present seem inexplicable.
>
> (Joly and Rutherford, 1913, p. 657)

Three years later Joly measured halos in younger rocks from the Vosges (Joly, 1917), and later still he measured the radii of rare halos in the Tertiary granites from the Mourne Mountains. The radii of the latter were 7% smaller than those of the Leinster Granite, which themselves were 10% smaller than those he had recognised in Archaean rocks (Joly, 1923b). Joly concluded in triumph: ''t would seem as if we might determine a geological chronology on the dimensions of these halo-rings!' (Joly, 1922a).

No sooner had he come to this exciting conclusion than he discovered small halos in Archaean rocks from Norway. His first specimens of this material had been lost when his laboratory at Trinity College was occupied by troops during the Easter Rebellion in 1916 (Joly, 1920). Six years later, he received fresh material and discovered the small halos which he concluded represented full decay of a radioactive element that was no longer present in the Earth's crust. He called this new radioactive element 'hibernium' after his homeland (Joly, 1922b) it was later found to be samarium.

In 1922, Joly reiterated his contention that radioactive decay rates were not constant throughout geological history. This was partially based on his observations of halos whose characteristics were not consistent. In particular, the innermost rings produced by uranium in some halos were not consistent with the known ionisation curves of uranium alpha particles. He suggested that these rings were caused by the faster decay of uranium in the past, or by the decay of a uranium isotope that was no longer present. The fact that thorium halos of all ages were constant in size was also difficult to explain (Joly, 1922a). Holmes (1926) argued that the inconsistencies of the uranium halos were due not to time but to other factors, including the presence of the recently discovered rare isotope, actinium. Nevertheless, he accepted that Joly's scheme of correlating halo radii with time would eventually give a scale against which the ages of other halos could be determined.

Subsequently, however, Joly's halo data were examined, and the accuracy of his measurements was questioned by Kerr-Lawson (1927) who was unable to detect the differences in radii that Joly had claimed. Nevertheless, his imaginative attempt to employ pleochroic halos as a measure for geological time was noteworthy. Joly's objections to dates derived by radiometric methods were also rejected by the late 1920s when further information about isotopes, atomic weights, and other radioactive elements became available (e.g. Holmes and Lawson, 1927). Holmes, who was at the forefront of radiometric research (Lewis, 2001), argued that, at worst, errors of a few per cent might be attributed to the unranium-lead methodology (Holmes, 1926).

Earth's surface history and thermal cycles

Joly's final contribution to the geochronological debate came about from his interest in isostasy, the internal heat of the Earth, and its signature on the surface of the Earth. Kelvin's ideas on a cooling Earth were long dismissed, and it had been recognised by Strutt and Joly that the Earth had an internal heat source, resulting from decay of the radioactive elements it contained. Joly's pioneering suggestion was that the internal heat of the Earth built up over a period of time (Joly, 1909c).

By 1923, Joly had developed a theory in which he explained that the surface features seen on the Earth were formed as a direct result of its own internal heat source. This theory subsequently became known as the theory of 'thermal cycles' (Joly, 1928). The heat source was responsible for the melting of the basaltic crustal horizons (termed the basaltic magmatic layer by Joly) that lay beneath the continental crust. The continents, which he described as 'granitic scum' (Joly, 1925, p. 176), rested directly in isostatic balance on a basaltic magmatic layer that was up to 70 miles in thickness, and which enclosed the whole globe. The water that made up the oceans simply filled the voids between the continents. In a thermal cycle, he explained, this basaltic magmatic layer melted periodically due to the build-up of heat generated by radioactive decay. Melting led to changes in the volume of this horizon beneath the continents, which in turn generated tidal currents in the basaltic magma. Joly argued that such a change in volume could have resulted in an increase in the diameter of the Earth, and that the continents would have been isostatically buoyed up. This effect would have been negated by the resultant sinking of the continental masses due to the decreased density of the molten layers, and consequently, there would have been widespread transgressive events.

What is interesting is what Joly speculated happened on cooling of the molten basaltic magmatic layer. Heat was lost as it migrated from beneath the continental areas to oceanic regions, from where it was largely lost into the oceans (Joly, 1923a). Joly argued that the cooling of the melted level caused it to recrystallise and shrink, and that continents would return through a series of vertical movements to their former isostatic levels. He pointed out, quite correctly, that vertical movements were often associated with lateral or horizontal Earth movements. Both, he argued, were generated by the same thermal processes outlined above. But how did he explain the orogenesis that was also a feature of these complex processes? He postulated that the most lateral movement occurred when the basaltic level beneath the continents was in a molten state, at which time it would have had a similar density to the granitic continental masses above. Beneath the oceans thermal currents significantly reduced the thickness of the basaltic magmatic layer, which he deduced would have cracked, and which would have seen the injection of new basalt. The area of the ocean floor thus increased significantly, and this increase in area resulted in the generation of compressive forces which forced the oceanic floor to press against the margins of the continental crust. This initiated orogenesis or mountain building. Joly estimated that these orogenic events or 'revolutions', as they were called, took place once every 50 million years or so. He tabulated five or six revolutions of different ages which he correlated with various mountain orogenic belts. These included the Caledonian, Appalachian and the Alpine 'revolutions'. From the evidence of global tectonics, Joly considered that the Earth was 300 million years old.

Many of his ideas on global tectonics were articulated in his book *The Surface History of the Earth* (Joly, 1925, 1930) and in his papers on thermal cycles (Joly, 1928) and the *Earth's Surface Structure* (Joly and Poole, 1927). Naturally, he entered the debate on continental drift and argued that there was evidence that the continents drifted westwards. This delighted some Irish wags, who highlighted in the popular press this scientific proof that Ireland was moving further away from England.

Joly's ideas on global tectonics were typically complex and imaginative, and certainly deserve fuller examination and assessment this is beyond the scope of this present paper. Greene (1982) has discussed the evolution of tectonic theory at length, from forerunners such as Elie de Beaumont and Eduard Suess, to those later geologists that included Joly, who worked on what Greene termed 'the fourth global tectonics'. Oreskes (1999) also analyses in depth Joly's contribution to the theory of continental drift.

Conclusion

As early as 1922 Joly noted that the expression the 'age of the Earth' was ambiguous (Joly, 1922a), and it can now be seen that Joly's own position on this matter was also ambiguous. In 1930, in the second edition of his *Surface History of the Earth,* published three years before his death, he wrote: 'the age of the Earth is not the same as geological time' (Joly, 1930). It is my contention that Joly understood 'age of the Earth' to mean the time since the onset of denudation and biological processes, while 'geological time' included older Archaean times. In essence, Joly was taking a philosophical stance which allowed his low age estimates to stand as correct estimates for the age of the Earth, yet at the same time he did not reject out of hand the accepted estimates of Holmes and others, derived by radiometric means, which gave a longer measure of geological time. Nevertheless, he remained sceptical of dates determined by radiometric methods.

In the light of his work on radioactivity, it is somewhat surprising that Joly did not accept that his sodium method yielded erroneously low age estimates, or acknowledge that what it measured was not the age of the Earth, but an estimate of the age of the oceans. Although it is clear today that the oceans are older than Joly's method might suggest, his sodium method actually gives a good measure of the residence time of sodium in the oceans, which is approximately 260 million years (Mittlefehldt, 1999). Up until his death Joly continued to hold the view that 300 million years was a good estimate of the age of the Earth.

Joly's age methodologies and breadth of research in geochemistry, mineralogy, and radioactivity made a major contribution to the debate on the age of the Earth played out at the end of the nineteenth and beginning of the twentieth centuries. His sodium method held centre stage among geologists and geochemists for some years and stimulated much debate on the subject of the age of the Earth. Additionally, Joly's unique attempts to employ pleochroic halo radii as measures for geological time were extremely valuable and, in the light of the exciting avenues opening up in the field of radioactive research in the first decades of the 1900s, may well have appeared to have the distinct potential to estimate the actual age of the Earth. Even though it was subsequently demonstrated that the ratio between original radioactive elements and their decay derivatives yielded better estimates of the age of the Earth, Joly's many faceted investigations on the subject helped to stimulate research that ultimately led others to the geochronological methodologies in vogue today.

Joly spent all of his professional life working in Trinity College, Dublin, during which time he wrote 269 scientific papers and several books, including an autobiography. He was the recipient of many awards and had a high international reputation. He was a Fellow of the Royal Society which awarded him the Royal Medal in 1910, and he received the Boyle Medal of the Royal Dublin Society in 1911 and the Murchison Medal of the Geological Society of London in 1923. He became a Fellow of Trinity College in 1919, the first person in 250 years who did not have to pass a rigorous examination, and was president of the Royal Dublin

Society (1929–1932). He was conferred with the honorary degrees of Doctorate of Literature from the University of Michigan, and Doctorate of Science from the University of Cambridge and the National University of Ireland. Forty years after his death, he was honoured by having a crater on Mars named after him (Batson and Russell, 1995), which is appropriate given Joly's work on the nature and origin of Martian 'canals' (Joly, 1897).

He was by all accounts a very popular man, loved and respected by many. On his death many friends contributed to a memorial fund that is still used to promote an annual lecture on a geological theme, which is usually given by a leading foreign authority. In addition, his name and memory are perpetuated by the Joly Geological Society, the student geological association of Trinity College, Dublin, founded in 1960.

18

GRENVILLE ARTHUR JAMES COLE (1859–1924): THE CYCLING GEOLOGIST

In P.N. Wyse Jackson (ed.) *Four Centuries of Geological Travel: The Search for Knowledge on Foot, Bicycle, Sledge and Camel.* Geological Society, London, Special Publications, 287 (2007), 135–147.

--

To those who laid from west to east
The clear white road for man and beast;
To those who in our age of steel
Set the true frame and shaped the wheel;
To those whose greeting calls again
From the brown acres of the plain,
Or through the pathways of the firs,
Our equal friends and wayfarers;
To him who at each sunset post
Held wide the door, our welcome host;
To these, on many a country-side,
Our thanks be given, as we ride.

(Cole and Cole, 1902, Dedication)

Since the seventeenth century, men and women have been travelling parts of the globe in search of rocks and geological structures that could elucidate something of the Earth's history. They have travelled by many different means: James Hutton, William Buckland, and early Geological Survey mappers preferred the use of horses, Roderick Impey Murchison and his wife travelled by horse-drawn coach (Kölbl-Ebert, 2007), whereas the fossil fish-collectors William Willoughby Cole, the third Earl of Enniskillen (1807–1886) and Sir Philip de Malpas Grey-Egerton (1806–1881) were among those gentlemen 'geologists' who made Grand Tours of

DOI: 10.4324/9781003475552-22

Europe during the eighteenth century. Rail travel prevailed in the early twentieth century but has now been overtaken by the motorcar. In remote areas such as Africa, geologists have resorted to canoes, or their feet. At one point towards the end of the Victorian era, cycling was very popular.

The earliest wheeled vehicles were in use in Mesopotamia around 7,000 years ago, but it was not until late in the eighteenth century that self-propelled two-wheeled machines first appeared. These were the so-called 'hobby-horses', which were awkward and heavy, consisting of two large wooden wheels and with a saddle fitted to a rigid crossbeam. They could not be steered easily and were propelled by the rider running his feet along the ground in a 'Fred Flintstone' manner. These machines were particularly popular with the aristocracy of the late 1700s and early 1800s. Not content with this self-powered propulsion, early nineteenth-century inventors toyed with pedals, levers, and chains in an attempt to perfect efficient mechanical movement. The first recognised bicycle was built by a Scot, Kirkpatrick Macmillan (1810–1878), in 1839. A replica of his machine may be seen in the Science Museum, London. From this start evolved the hard riding 'boneshakers', which were large and heavy, like the earlier 'hobby-horses', but which were easier to ride. They came in many shapes and sizes: although two-wheeled versions were the most prevalent, three- or even four-wheeled machines were not unheard of. The introduction of the boneshaker had important social implications. They were relatively cheap at £10 in the 1860s and so bicycling became available to the general public. Two further advances were yet to come before modern machines took shape: the first was the advent of the light triangular-framed machines typified by the 'ordinary' (or 'penny-farthing'), which was followed, in the 1880s by the chain-driven 'safety'. The second advance was the invention of the pneumatic tyre by John Dunlop (1840–1921) in 1888. This tyre is now on display in the Royal Scottish Museum. Dunlop was a Scottish veterinary surgeon working in Belfast. Soon after 1888, he moved to Dublin where his pneumatic tyres were manufactured (Cooke, undated). At last, bicycling had become a comfortable means of transport and available to all.

Where do geologists fit into all this? Bicycling fever swept through Victorian Britain and many touring clubs, of which the Cyclists' Touring Club was the premier, were established. Such enthusiasm slowly spread to the United States so that by the early years of the twentieth century, the pastime was as popular as in Britain (Smith, 1972). Any volume of material was provided for cyclists: publications included periodicals such as *The Cycling Mercury*, *The Cyclist*, and *The Wheel World*, the latter edited in part by Henry Sturmey noted for his gear set-up; maps on which the most suitable routes were marked. These included Mason and Payne's *Cycling Map of the British Isles with Parts of France, Belgium and Holland*, scale *c.* 1:1,584,000, London, 1888, and the series of maps produced by George Washington Bacon and Co., London between 1885 and 1923 that included *Bacon's Cycling Road Map: Ireland*, scale 1: 253 440, London, 1900; guide books complete with profiles of routes showing gradients (Inglis, 1896, 1897, 1908);

manuals on how to repair tricycles and bicycles; and listings of the cycling clubs and hotels that welcomed touring parties (Spencer, 1884). There was considerable debate about the merits or otherwise of cycling for women. This activity for females was promoted by Miss Erskine and others (Erskine, 1884, 1896, 1897; Graves *et al.*, 1898) but its dangers were pointed out by others, including Mr Ryley (1899). Catalogues advertised clothing that was both suitable for lady cyclists and which retained their modesty.

It is hardly surprising that some geologists should become committed cyclists. The bicycle widened the field area for research and allowed the rider to become personally entwined with the landscape, a feeling that could never be adequately attained by rail travel. In the United States, Helen Margaret Duncan (1910–1971) a specialist on Bryozoa and Geological Survey palaeontologist from 1942 until her death, was apparently a keen cyclist (Flower and Berdan, 1977). In the British Isles, the Geologists' Association, a group of professional and amateur members, undertook a number of excursions by bicycle in the 1890s and early 1900s (Green, 1989; Robinson, 1990). Amongst this group, cycling as a mode of transport was championed by the Rev. John Frederick Blake (1839–1906) formerly a Professor of Geology at Nottingham and by Horace Woollston Monckton (1857–1931). After a short time, their enthusiasm for the wheel fizzled out. However, the individuals Sydney Savory Buckman (1860–1929) who was an ammonitologist, and Grenville Arthur James Cole (1859–1924; Figure 18.1) an igneous and metamorphic petrologist, were both committed cyclists (Wyse Jackson, 1991). In eighteen months from the date of his purchase of his first pneumatic-tyred bicycle, Buckman covered 3,000 miles, principally engaged in examining various Mesozoic successions in Normandy and around his home in the Cotswolds (Buckman, 1898). Buckman, together with his wife, established the Western Rational Dress Club that advocated that women wore a knickerbocker costume. In 1900, he seriously damaged his health through over-exertion in the saddle (Torrens, 2004b).

Of these cyclists, it is the little-known Cole who has left most written and other material relating to his cycling activities and from this, it may be seen how cycling influenced his geological and geographical work.

Biographical notes

Grenville Arthur James Cole was born in London on the 21 October 1859, the second son of John Jenkins Cole, architect to the London Stock Exchange (Wyse Jackson, 1989). He was educated at the City of London School and, in his final year began to attend the lectures of Professor John Wesley Judd (1840–1916) at the Royal School of Mines. These captivated Cole, and although he was not destined to enter university, he managed to embark on an academic career in geology. Judd recognised Cole's potential and in 1878 appointed him as demonstrator to the school, a position he held for 12 years. During this period from 1886 until 1890, Cole was also Head of the Geology Department at Bedford

College for Ladies. Soon after he left London, a student geological society was established and named in his honour. Bedford College was later incorporated into the Royal Holloway, part of the University of London, which appropriately still awards an annual Grenville Cole Travel Bursary which funds a student to engage in research.

In 1890, Cole was appointed as Professor of Geology and Mineralogy at the Royal College of Science for Ireland in Dublin (Kelham, 1967), only a five-minute cycle from Dunlop's tyre factory, in succession to Edward Hull, and he retained this position until his death thirty-four years later. This institution had been established to provide education in science as applied to agriculture and industry and also training for teachers destined for the technical and secondary schools. Cole was later appointed as Director of the Geological Survey of Ireland in 1905 and was active in most of the societies that promoted science and natural history in Dublin. These were the Royal Dublin Society, the Royal Irish Academy, the Dublin Naturalists' Field Club, the Dublin Microscopical Club, and the Irish Geographical Association. He was a member of a close-knit group of scientists who used to meet for buttered toast and tea in The Farm Produce Depot and Refreshment Rooms adjacent to Trinity College, Dublin, run by the Misses Gardiner. Cole and the group presented their hosts with a series of marble-topped tables which he described in quaintly named paper 'Geologists at the Luncheon Table' (Cole, 1894a).

In 1896, Cole married one of his first students in Ireland, Blanche Vernon (Figure 18.1), the daughter of Colonel J.E. Vernon of Clontarf Castle, a wealthy Dublin landowner, and the couple had one son, Vernon, who later joined the East African Civil Service but subsequently practised as a medical doctor. Blanche, like her husband, was an enthusiastic cyclist, and together they travelled extensively in Europe, as well as in Algeria and Tunisia. According to the botanist Robert Lloyd Praeger, Cole's best man, after their wedding ceremony the happy couple mounted their bicycles, cycled down the aisle and embarked on their honeymoon! (Praeger, 1941, p. 272).

Towards the end of his life, rheumatoid arthritis progressively crippled Cole, but he remained characteristically cheerful and active. Grenville Cole died on 20 April 1924 at the age of sixty-four, and Blanche only survived him by three years, dying at the same age on 17 November 1927. They are buried together in Dean's Grange Cemetery in south County Dublin.

During his career, Cole received numerous awards. He was awarded the Murchison Medal of the Geological Society of London in 1909, was a President of Section C (Geology) at the 1915 British Association for the Advancement of Science, was elected a Fellow of the Royal Society in 1917, and received the degree of Doctor of Science, his only degree, from Queen's University, Belfast.

Grenville Cole was a small man who stood little over five feet tall, but who had endless energy. He would frequently walk fourteen miles to and from work each day, and he expected his students to work hard. He was a superb lecturer, field geologist, and teacher, who had a captivating wit and a sparkle in his eyes.

Figure 18.1 Grenville and Blanche Cole (sitting on ground) with a friend (standing at left) pausing during a rain shower while on a cycling trip somewhere in Ireland, *c.* 1900 (Photograph by John Joly, now in the Geological Museum, Trinity College Dublin).

Cole's geological work

At the Royal School of Mines, Judd advocated the use of practical methods (field studies, excursions and practical classes) to complement lectures. There was a shift in emphasis from the lecture theatre to the laboratory, so that by the turn of the century, practical classes, demonstrations, and field classes were soon practised in many university departments in the British Isles and on the Continent. Cole was responsible for much of the planning and teaching of the practical element of the geological courses. The job of demonstrator also allowed him time to pursue extensive research interests.

After 1890, his duties at the Royal College of Science in Dublin were to organise and deliver lectures to various groups: agricultural students, science students, and student-teachers. In an average year, he or his assistant gave approximately eighty lectures. His assistants at one time or another included Louis Bouvier Smyth (1883–1952), later Professor of Geology and Mineralogy at Trinity College and Isaac Swain (1874–1963), later Professor of Geology and Geography at the University College Cork.

Cole's career may be divided into two periods where the thrust of his research was quite divergent. This first period was up until 1895 during which time he worked largely on petrology and the illustration of textures and relationships within igneous and metamorphic rocks; in 1889, the Geological Society of London awarded him the Murchison Fund for this work. At this time, he travelled extensively in Britain, collecting Scottish basalts and Welsh volcanic rocks.

Later, he became more involved in the promotion of fieldwork and the study of physical geography. He advocated the teaching of geography in universities and helped raise the profile of the subject (Herries Davies, 1977b). He was a prolific author of both academic and popular geological papers and books and has over 500 papers to his name (Wyse Jackson, 1989).

The cycling geologist

Inspired by Judd's teaching, Cole began to comprehend the importance of outdoor field studies in geology. Perhaps in response to this, he began cycling. In the preface of *Open-Air-Studies: an Introduction to Geology Out-of-Doors* which resulted from first-hand field experience, he wrote 'Now-a-days, when cycling is so frequent and so free a means of travel, the geography and geology of nearer Europe have become keen realities to many of us…' (Cole 1895a). Field studies were an aspect of the science that preoccupied his thoughts and many of his publications for the rest of his life. Cole was an avid cyclist—this cannot be emphasised enough—and this allowed him to undertake considerable geological and geographical fieldwork.

Apart from his cycling tours Cole travelled widely. He visited Canada in 1897 and South Africa in 1905 to attend the meetings of the British Association for the Advancement of Science and regularly attended when it met in Britain.

Reconstructing Cole's travels

Using various lines of evidence, I have established the routes of many of his cycling tours. Of most value in this exercise were his two travel books *The Gypsy Road* (Figure 18.2) (Cole, 1894c) and *As We Ride* co-authored with his wife Blanche (Cole and Cole, 1902), as well as his geological textbooks illustrated with his own photographs and any scientific papers similarly illustrated. These include papers on Algeria (Cole, 1918a, b) and Spitsbergen (Cole, 1911b), which he visited in 1911 during an International Geological Congress meeting. An unexpected bonus, and source of valuable information about some of his European and British travels, was his collection of rocks found in the basement of the Museum Building in Trinity College, Dublin. These were still wrapped in scraps of annotated newspaper, which gave some indication of when and from where he had collected the specimens. Fortunately, Cole had very distinctive, clear handwriting, and he habitually wrote in black ink, and it is not possible to mix this up with that of other hands. His student field guides produced for the student field trips, together with his manuscript map of Ireland that is in the Geological Museum, Trinity College Dublin, provided valuable information. On this dissected map, he had marked many routes travelled by him in red ink and added some information about hotels, hostels, and local geological stops. Grenville and Blanche also kept a large map of Europe in their home and this was marked with the routes of their cycle tours. After their deaths, their son sold the house, and the new owners, the Bewley family of the noted Dublin café, kept the map in an outside shed before it was regrettably discarded.

Travels around continental Europe

The Gypsy Road

By 1894, Cole had completed journeys on a Humber-Beeston tricycle through Italy and Bohemia in tandem with a companion Gerard W. Butler who rode a penny-farthing. Cole had owned his machine since 1884 and had used it on several trips. He was very proud of this tricycle and said that ownership of it opened doors throughout Europe. He had dined with the famous Hunnia Cycling Club of Pest and frequently was joined by local cycling groups. Arising from these, trips came papers on volcanic rocks of Lipari (Cole and Butler, 1892), Italy, the Vosges, Germany (Cole, 1887), and parts of France (Cole and Gregory, 1890).

In the summer of 1892, he travelled to Hungary to compare the volcanic sequences found there with those Irish examples already familiar to him (Cole, 1892b, 1894b). Following this journey, he wrote *The Gypsy Road*, a highly entertaining, perceptive, and lucid travelogue and social commentary (Cole, 1894c), which was described in *The Times* edition of 12th May 1894 as being a 'breezy narrative of a cycling tour in out-of-the-way regions'. It recounts a journey of 1,055 miles westwards from Krakow to Coblentz, through what is now Poland, the Czech Republic, Slovakia and Germany, which Cole and Butler, guided by *Bradshaw's Map of Europe*, completed in 38 days in 1892! In fact, not content with travelling in one direction, Cole actually cycled from the town of Zolyom over the Szturecz Mountains with another friend before rendezvousing with Butler in Krakow.

Butler is named throughout the narrative the 'Intellectual Observer' and comes across as being a man of quiet wit and perception. In response to a question from a native as if whether there were any mathematicians in England, he replied 'A few perhaps at Cambridge' (Cole, 1894c, p. 127). Little is known of Butler and his background; he was a zoologist, enjoyed painting, held a B.A. degree, was a Fellow of the Geological Society, who studied and published on snakes and lizards (Butler, 1889). He also visited Pantellaria, south of Sicily and described its 1891 active volcanic activity (Butler, 1892; Butler and Perry, 1892). Cole, on the other hand, is the extrovert of the pairing.

They did not find everything plain cycling. The local police in places had some reservations with their visas, they were taken for travelling salesmen in other towns, horses were scared of their machines and they had to contend with bed bugs or being chased by geese or by fifty children—one wonders which is worse.

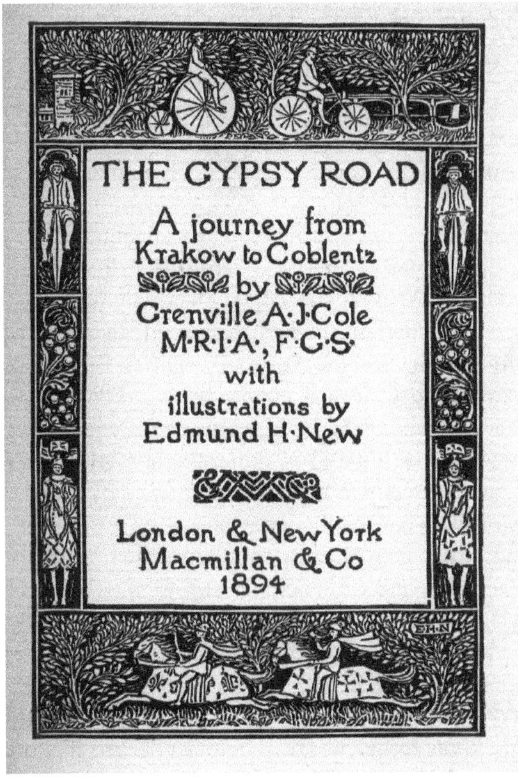

Figure 18.2 Title page of Grenville Cole's *The Gypsy Road* 1894 (Author's collection).

They coped with high temperatures in Bohemia where by mid-day it was 90°F. On 19 August between noon and 7.30 pm, they consumed in total one large flask of sparkling water, two cups of tea, three glasses of water, one hot coffee, and two iced coffees, and ate only three small rolls and one-half slice of Hausbrod. This Cole stated 'express[es] the situation better than any thermometric record' (Cole 1894c, pp. 135–136).

In one Czech village, they arrived at the inn and asked for a room for the night. Unfortunately, they enquired in German and were refused. Somewhat alarmed, as it was approaching dusk and the next village was several miles away, they went to the police station where they received the same cold-shoulder treatment. When the villagers realised that they were English and not German they were immediately received with open arms. The best room in the hotel was made available to them and their conveyances were parked in another guest room. On asking why this transformation in attitude to them had taken place, they were told that although the locals could speak German, they certainly did not speak it to Germans! Elsewhere they encountered this antipathy to Germans; they were asked 'Are you Christians? Or are you Germans, perhaps?' (Cole, 1894c, p. 80).

Geologically, this trip was interesting and diverse. From Poland, they travelled south, passing through Myslenice, which had been visited in 1818 by the mineralogist Françoise Sulpice Beudant (1787–1850) who recounted the trip in a four-volume work *Voyage minéralogique et géologique, en Hongrie, pendant l'annee* 1818 (Beudant, 1822), then they ventured into the Hungarian alluvial plains, before crossing the limestones of the Tatra Mountains. They then entered the region of the north-eastern fore-Alps and the Hungarian Alps where they cycled over Cenozoic sediments, Mesozoic limestones and occasionally encountered crystalline basement rocks. At Garam, they swam in the meandering river before cycling through the rolling volcanic countryside in the moonlight. Near Hajnik, they visited the ancient gold mining region of Schemnitz, which had been renowned as being the most important centre for the metal in the Austro-Hungarian Empire. Nearby, the travellers collected obsidian and Cole was able to examine the local rhyolites in some detail. Thanks to an invitation by Dr Cseh, a government geologist, they were able to venture underground into the huge silver mine at Schöpferstollen where they walked through the workings for over two hours, accompanied by a mine manager and two small Slovakian boys who carried the lanterns.

They found the expansive Hungarian Plain, underlain with unconsolidated sediments, rather tedious, and Cole compared it with Salisbury Plain in England. In Hungary, they followed in the tracks of the earlier traveller Richard Bright (1789–1858) who has been described as being the 'Father of Nephrology', who had been there in 1815 (Bright, 1818). Once in Moravia they passed over Miocene conglomerates and noted the breathtaking local geomorphology: 'It is wonderful what amount of rock has to be cut away before you can make a decently artistic mountain'. In Prague they met with Antonín Fric (1832–1913), who worked in the Museum of the Czech Kingdom in Prague and was Director of Palaeontology there from 1880. He arranged for the travellers to visit the volcanic regions to the north of the city, and they spent a number of days examining the rocks around the value of Algersdorf and climbed various volcanic cones such as the Geltschberg where rare exposures of white phonolite could be seen. In Britain, Cole had seen the only exposure of this rock type at the Wolf Rock in Cornwall and rejoiced in 'the joy of riding in Bohemia from one crag to another, smiting the solid masses with the hammer, and seeing the lustrous flinty chips go skimming across the sunburnt grass' (Cole, 1894c, p. 131). In Bohemia, Cole was in his element: wonderful rocks and wonderful cycling. Near Karlsbad, they visited the volcanic cone of the Schladnigberg and traversed the deposits of Oligocene and Miocene brown coal near Dux, noting that some seams had been burning for at least five years and could not be extinguished, before passing out of Bohemia and into Bavaria.

By now, their journey was nearing its end, and Cole having seen the volcanic deposits of Hungary could relax somewhat. He drank schnapps in a tavern in Horgenau, having been recommended to do so on medical advice by some of the elderly men of the village. He declared that his 'union with this simple country life seemed complete' (Cole, 1894c, p. 159). However, not much further down the road in the village of Werdorf the journey nearly came to an abrupt end.

Having parked their machines, Cole and Butler were relaxing when the tricycle started to freewheel downhill. It collided with a low wall, somersaulted twice and landed in a cabbage patch. The axle was bent and it seemed that the tricycle would have to be brought to Wetzler to be repaired. The town would have interested Cole the petrologist as it was there that Leitz petrological microscopes and other optical instruments were and continue to be manufactured. Fortunately, the local blacksmith, aided by Cole and Butler, managed to take the wheel bearings apart, straighten the central axle and reassemble the mechanism. During this operation, the forge was visited by a man who wished to have his horse reshod. The blacksmith dismissed him, telling him that he now ran a *Velocipedenfabrik*! The wheels ran smoothly afterwards 'and were good for another thousand miles' (Cole, 1894c, p. 161).

Finally, the pair of cyclists reached Coblentz and Cole likened them to Knights-Errant (Cole, 1894c, p. 166); this comparison is a theme that crops up every so often in his writings. His book *The Gypsy Road* is important as it is one of the earliest travelogues of a bicycling trip and it gives an insight into eastern European lifestyles of the period.

As We Ride

In 1902, Blanche and Grenville Cole published a slim book of essays entitled *As We Ride* (Cole and Cole, 1902), which gives accounts of a number of their trips through France, Germany, Poland, and the Balkans. This volume, now very rare, was published for the benefit of the Royal City of Dublin Hospital and it provides a glimpse of pre-First World War Europe.

In addition to containing geological insights, the essays are generally observations of the social interactions of local inhabitants and the towns through which the couple passed. One senses that Grenville had moved in his mind from hard rock geology to focus more on geomorphology. Indeed, many of his publications from this period relate to the subject and its expression in Europe and to the interactions of man and his landscape (Cole, 1912a, b).

Of the eight essays, half were written by Grenville and half by Blanche. She certainly did not mince her words. Writing about a visit to the border town of Mlawa in eastern Poland in the aptly titled piece 'Defeat', she says of the landlady of the only hotel in the town:

> She was villainously unclean, and so vacant besides that I wondered whether she were deaf, or idiotic, or both. But the length of her bill next morning in proportion to the nature of our entertainment settled this point;... she was by no means idiotic-far from it.
>
> (B. Cole in Cole and Cole, 1902, pp. 100–101)

Rather than continue into Russia the couple retrieved their baggage and returned to Warsaw on the first available train. In 'Son of Mars', Blanche pins her political leanings to the mast: ‘War is an undesirable occupation and seldom profitable; but if one must fight, please give me the French to fight with, and let the Germans fight someone else’ (B. Cole in Cole and Cole, 1902, p. 77). Twelve years later, her sentiments counted for little.

Some of these journeys inspired Grenville to write poetry: *The Lost City* published in 1911 is his only piece of published work of this genre, and it recalls the history and evolution of a Templar hospice in the Causse de Larzac in Aveyron in south-east France (Cole, 1911a). The Coles were clearly taken with the Balkans which they visited on perhaps two occasions. Orahova, a small village now in Bosnia-Herzegovina, provided them with the title of a chapter in their book, but also and more significantly the name of their house in the fashionable district of Carrickmines in south County Dublin which they had purchased in 1902. The house is still standing and was renamed Glenheather by a subsequent owner.

A number of these excursions were recalled later by Cole in public lectures, including that on ‘The Fringe of the Balkans’ delivered at the Royal Institution in London on Monday 8 February 1904 (announced in *The Times* 30 January 1904). Like *The Gypsy Road*, *As We Ride* is a fascinating commentary on a period long gone and a reminder of the remarkable energy and close bond of Cole and his wife.

Travels in Ireland

General excursions

In 1884, Cole paid a visit to Ireland, but other than knowing it was to examine the geology of the island, and that it was one of the first excursions on his Humber-Beeston tricycle, nothing is known of the localities that he visited. Soon after his arrival in Dublin in 1890 on his appointment to the Royal College of Science, Cole rapidly set out to reacquaint himself with the geology of Ireland and did so largely by bicycle. In 1892, just two years after his arrival in Ireland, he penned a series of illustrated articles on the geology and scenery of Dublin that appeared in *The Irish Naturalist* (Cole, 1892a). He was then commissioned by the owners of the North Eastern Railway in 1895, to write a similar booklet outlining the geology and scenery as seen from their railway carriages. This booklet, to accompany the photographs of the Belfast photographer Robert John Welch (1859–1936), was thus the first of its kind in Ireland (Cole 1895b). Welch was the foremost natural history photographer of his time in Ireland and his images illustrated many scientific papers including those authored by himself. He was an excellent conchologist. He was commissioned to photograph the *Titanic* and its sister ships while they were being built in Belfast, where he had his studio (Evans and Turner, 1977).

To satisfy his fascination with volcanic rocks, he sought out intrusions of rhyolite and headed north to north-east Antrim (Cole, 1896). He then examined the volcanic mass of Slieve Gallion (Cole, 1897). Later, he carried out research on the Dalradian metamorphic belt in Donegal, produced accounts of the Donegal granite and its orbicular structures (Cole, 1916), which is now sadly damaged by coring, and reports on the intrusions of Tyrone and Londonderry. Later work embraced geomorphology, although he continued to write on igneous and petrological topics. One of the most important papers of his later years was that which described the course of the River Liffey, the river that bisects the centre of Dublin (Cole, 1912c). He also put his bicycle to patriotic use: it has been reported that throughout the 1914–1918 war, he rode the streets of Dublin as an army dispatch rider (Herries Davies, 1995).

Leading student field excursions

Every year between 1903 and 1908, Cole took a group of his second-year agricultural students, or 'scholars' as he called them, on a week-long trip around Ireland to study geology and agricultural botany. This excursion was often held in the month of May and much of the travel was undertaken by bicycle or by train. These excursions first started in 1903 following the sanction of the scheme and agreement to fund it was given on 16 April by J.D. Daly of the Department of Agriculture and Technical Instruction, a government office responsible for the college. The cost of the first trip, which included hotels and train travel, was £5/9/11 each (approximately £340 today). It did not include the price of lunch, which was taken in the field, and which each person in the party was expected to provide for himself.

Cole expected his students to cope with the rigours of cycling field trips. As can be seen from the attractive excursion booklets that were provided for the trips, they were quite strenuous. Booklets were produced for the 1903, 1904, 1905, 1906, and 1908 trips and are similar in style, being approximately twelve pages long and A5 in size. They were mass-produced on a copying machine in black, blue, or purple ink and were illustrated with geological cross-sections drawn by Cole and with pasted-in portions of maps cut from the recently reissued *Murray's Handbook for Ireland* (Cooke, 1902). Copies and associated correspondence quoted here are in the archives of University College Dublin. These booklets documented the routes to be covered and the features to be examined on the trips. When planning the trip Cole wrote to Edward Stanford, publisher of the maps in Murray, asking him to furnish unbound sheets of these maps. Stanford's response dated 21 April 1903 indicated that he had only printed enough sheets for the binding of the volume and that no additional copies were available at that time. A subsequent postcard from Stanford to Cole, dated 23 June 1903 said that he would bear this request in mind should a further printing of the *Atlas* be needed.

Taking the 1903 trip as a typical example, we find that on 18 May, the group met at Broadstone Railway Station in Dublin at 6.45 am. Following the departure of the train at 7.00 am, they took breakfast in the second class car. Five and a half hours and approximately 140 miles later, the party disembarked at the small town of Recess west of Galway having crossed over the Central Plain of Ireland. Rather than gaze idly out the window, the students were expected to take notes on the features that they saw. From Recess, they bicycled 17 miles to Leenane where they examined Dalradian rocks and the vegetation. The next morning, they cycled 30 miles over the Silurian conglomerates and silts and Carboniferous rocks of the Mayo district. On 20 May, an early morning train carried the group north-eastwards and, following a break from rail travel during which they cycled 21 miles and examined geological and botanical features of interest, they ended up in the town of Omagh. The next day they first visited Pomeroy which had yielded important Silurian fossils first described by Joseph Ellison Portlock (1794–1864) in his monumental volume on the geology of Londonderry (Portlock, 1843; Tunnicliff, 1980). After a 26-mile cycle over old gneissic terrain and undulating eskers, they passed Slieve Gallion and by evening had almost reached the edge of the Antrim Plateau, which geologically comprises a series of Tertiary basaltic flows. The next day, following their usual early morning rail departure, they reached Kilroot and visited an underground mine from which Triassic salt was being exploited; it is still mined and is mainly used in industry and by local authorities to grit the roads after a snowfall or heavy frost. This visit was followed by another 21-mile cycle into Belfast via the overlying basalt sequences exposed at Cave Hill having also examined en route the Triassic and Cretaceous sediments protected beneath the younger basalt. On the final day of this first excursion, 23 May, the students and Cole left Newcastle, County Down at 8.00 am and by 9.20 am had already climbed Slieve Donard, a nearby mountain. They then cycled to the small fishing port of Greencastle, loaded their bicycles on to a ferry and sailed across Carlingford Lough to Greenore. From there, they took a train to Dublin and arrived in the capital at 9.10 pm.

In 1904, Cole's party of two staff members and twelve scholars visited counties Tipperary and Limerick in Munster before travelling through western and northern Ireland. The following year the group had to cycle to Louisburg and climb Croagh Patrick in County Mayo, before heading south-east to the Knockmealdown Mountains in County Waterford. The cost per student for the 1905 trip was £6/7/5 excluding lunches and miscellaneous fees. The following year, in 1906, the agricultural students were joined by a number of third-year natural science teachers in training and they travelled through Ireland in a double loop centred on Dublin. Following his return, Cole received a letter from T.S. Gill of the sanctioning body in which he was praised for the success of the tour:

> I am glad to think that the tour was worked so satisfactorily and economically. The little hand-book is very attractive, even to a scientific ignoramus. I congratulate you and your fellow-teachers on the success of this piece of educational work, and I hope the students appreciate the great advantages which they are receiving at your hands.

In 1907, he took his students to Tipperary again and then to see the sections in the basalt and older Mesozoic rocks in NE Ireland. He was unable to get maps for his booklet and so produced a series of handouts copied onto foolscap paper.

On 16 January 1908, he received a letter from Professor Campbell, Dean of the College, stating that the Department of Agriculture and Technical Instruction had informed him that no money would be made available to fund the field trips and that 'unless the Treasury pay… these excursions are at an end'. Cole pleaded his case to continue the excursions and even offered to use £10 from his laboratory budget to subsidise the excursions. He received permission to spend no more than £3/10/0 per head, and he managed to run a 6-day excursion in March of that year. However, funding was not forthcoming the following year and the 'long excursion' disappeared from the curriculum, much to his disappointment. This must have been a serious blow to Cole, particularly as he placed huge emphasis on the educational merits of fieldwork.

Cycling and its influence on Cole's research

Cole's early petrological work often depended on materials collected during his cycling trips. By 1897, the thrust of his work altered as he began to document the importance of geological structure on the geographical landscape and he contributed short articles on the geological/landscape theme to *Nature* and *Knowledge* (Cole, 1898a–f). He accompanied William Morris Davis (1850–1934) the American geomorphologist on a European tour in 1911 (Davis, 1912). Davis, it must be said, preferred train travel to pedal power, this is hardly surprising given the expanse of the North American continent. Cole summed up his thoughts in several later books: *The Growth of Europe* (1914) and *Ireland the Outpost* (1919):

> The study of its [Europe] fundamental structure in relation to the wayward actions of its overlords is for most of us the very keystone of geography.
>
> (Cole, 1919)

He continued this theme in his addresses to the Geographical Societies of Britain and Ireland (he was President of both in 1919) in which he stressed the importance that geography should be introduced as a university subject reiterating again the importance of geology on human distribution (Cole, 1921a, b; Herries Davies, 1977b).

Conclusion

Cole's scientific legacy lies in the pages of his published research output on many topics, but perhaps, of most significance on the volcanic rocks of Europe and on the foundations of its landscape. However, he also provided important social documentaries that are perhaps now of greater interest given the political changes in, and increasing awareness of Eastern Europe at the present time. His two travelogues give a valuable insight into a bygone period when the area was less turbulent.

Although Cole finally had to give up cycling late in life, due to increasing disability caused by arthritis, he never lost his love for it. In *Who's Who* for 1924, he added one poignant word to his usual entry to this annual listing of persons of influence: 'Recreation: **formerly** cycling especially as a means of travel' (Cole, 1924, p. 581). Sitting in Orahova in Carrickmines in Dublin, gazing at the large wall map, he would have retraced his peregrinations across Europe and North Africa. In the Epilogue to *As We Ride* he wrote:

> Up and down, through our own Atlantic isles, or across the untracked steppes of newer lands, we ride from dark to dawn, part in memory, part perhaps in dream…
>
> (G.A.J. Cole in Cole and Cole, 1902, p. 107)

19

ERRATICS, INTRUSIONS, AND GRAPTOLITES

In M. Mulvihill (ed.) *Lab coats and lace: the lives and legacies of inspiring Irish women scientists and pioneers.* WITS, 2009, pp. 48–59; 173–176.

Introduction

Geology has traditionally been associated with men, clambering over boulders and scaling rocky heights with a hammer in hand. Yet, women have played significant roles in geological and naturalist societies in Ireland and made important contributions to geological studies on the island (Higgs and Wyse Jackson, 2007). Many of these women came from the north of Ireland, perhaps reflecting that region's rich and complex geological heritage. Contributions to geological studies prior to the 1920s were predominantly made by amateurs associated with the various field clubs around the country, and three of the women featured here fall into this category. Our fourth subject, Doris Reynolds, was employed as an assistant at Queen's University, Belfast in the 1920s, but such university appointments of women were rare at that time. Indeed, even today Ireland's geological community remains generally male-dominated, although in recent years, the number of women taking geology degrees at university has grown significantly as has the number of academics.

For most of the twentieth century, however, few women held significant positions in the university or government sectors. The first women geologists were appointed to the Geological Survey of Ireland (GSI) in the late 1950s (Herries Davies, 1995) and the first woman lecturer in a geology department in Ireland in 1963 (Wyse Jackson, 1994c). These women did not progress far up the promotional ladder and most moved on within a few years. In Trinity College Dublin, even in the late 1970s, there were some who opposed the progression of women students to the honours programme, but this opposition was resisted. By 2000, there were women geologists on the academic staff of all Irish universities and the numbers of women employed as geologists in the Geological Survey Ireland and Geological Survey of Northern Ireland had increased. In 2014, Marie Cowan was appointed Director of the latter organisation.

DOI: 10.4324/9781003475552-23

Glacial erratics in the north of Ireland

In 1840, Louis Agassiz, a Swiss geologist, travelled through Britain and Ireland and announced, to the incredulity of many scientists, that he had found evidence of former glaciers throughout the countryside. The evidence included glacial erratics: in strict geological parlance, these are large boulders or rock, that had been picked up by the ice and carried some distance and came to be deposited on top of a different rock-type when the ice sheets melted. The term 'erratic' was also applied to smaller fragments of exotic material caught up in sand and gravel deposited by the ice. While Agassiz's theory of the former Ice Ages took some time to be generally accepted, by the 1870s, many scientists had begun to map the distribution of erratics. This was in vogue at the time—among other things, it helped to reveal the direction and route taken by the long-vanished ice—so much so that the British Association for the Advancement of Science (BAAS) established a committee to report on the subject, and between 1874 and 1908, it even granted the considerable sum of £157-16s-6d to support this work (Howarth, 1931).

In Belfast, during the 1890s, a group of amateur geologists became interested in the distribution of glacial erratics in the north of Ireland, keen to determine the direction of ice flow in Ulster. They formed the Geological Section of the oldest field club in Ireland—the Belfast Naturalists' Field Club (BNFC), founded in 1863 and still on the go. Its secretary and leading light for many years were Sydney Mary Thompson (1847–1923) (Figure 19.1), who had a keen interest in geology and was aided in her labours by another woman, Mary Katherine Andrews (1852–1914) (Creese and Creese, 2004; Wyse Jackson, 2019b).

Thompson was born at Whitehouse, Co. Antrim, one of several children of James Thompson, a Belfast linen merchant, and a niece of the naturalist William Thompson. Following her schooling, and three years in Dresden, Germany, she enrolled at the Belfast Government School of Art in 1870 where she gained several prizes (Newmann, 1993). She later trained in South Kensington and subsequently joined two groups of artists on her return home: the Belfast Ramblers' Sketching Club and the Belfast Art Society, becoming patron of the latter from 1921. With her sister, she shared a studio in their father's house at Macedon Point, north of Belfast. At the age of 53, Thompson married a Swiss artist Rodolphe Christen and was from then on styled 'Madame Christen'. Rodolphe was the seventh son of Melchior Christen and lived at St Imier, near Berne in Switzerland. Following their marriage, they travelled throughout continental Europe for two years before building their own house, which they called St Imier, in Ballater, Aberdeenshire, Scotland. Unfortunately, the couple enjoyed a brief marriage: Rodolphe died at home on 7 September 1906 aged only 47. Madame Christen herself died of heart failure in Landudno, on the north coast of Wales on 16 July 1923.

Mary Katherine Andrews was born in Belfast, one of six children and four daughters of Thomas Andrews (1813–1885), Queen's University Belfast's first Professor of Chemistry and its first vice-president (Davis, 2002), and his wife Jane (née Hardie). His major research was on ozone and other gases and he had

Figure 19.1 Sydney Mary Thompson, Madame Christen at Murlough Bay, Co. Antrim (Ulster Museum).

shown how they could be liquefied, a state that is exploited in modern refrigerators. For many years until her death Mary, his unmarried daughter, lived close to the university at 12, College Gardens. Little is known of her education, and nothing of why she developed her geological interests. She may have had a university education in geology, but even if she was self-taught she was confident enough to attend the Congress Auxiliary of the World's Columbian Exposition which took place in Chicago in August 1893.

This event, also known as the Chicago World's Fair, was part of the celebrations to mark the 400th anniversary of Christopher Columbus's discovery of the New World. There, with other women from North America and Britain she contributed to the Woman's Section of the Congress on Geology—the men met separately for their own convention (Bancroft, 1893). Short sessions were held throughout the week: Andrews gave a paper on 'Physical Geology', while Agnes Crane of Brighton read one on 'The evolution of the Brachiopoda', and Jane Donald of Carlisle spoke on 'Paleontology'. A list of the papers presented in the woman's convention is given in *American Naturalist* 1893, p. 1028. Donald was a noted expert on fossil gastropods (the class of animals that today includes snails and whelks) from the Palaeozoic era, some 540–250 million years ago (Wyse Jackson and Spencer Jones, 2007).

We can get a sense of Mary Andrews's character from the fact that, when she later became interested in the metamorphosed chalk found at Scawt Hill north of Larne, she did not publish her findings. Instead, she freely gave her rock specimens and observations to George Gough of Cirencester who utilised them in his assessment of the chalk and its metamorphism (Gough, 1907). That a woman would allow a man to publish her findings is typical for the time, but now is ridiculous and unjustifiable.

Sydney Mary Thompson was, from an early age, a keen student of glacial features. In 1894, she was elected a member of the Glacialists' Association and she also joined the Geologists' Association (both associations aspired to be international but drew their membership primarily from Britain and less so from Ireland and continental Europe). Thompson quickly began to make her mark as a geologist. In July 1895, she organised a visit to the region by the Geologists' Association, during which they enjoyed her hospitality with a visit to her father's house (McHenry, 1896). She and Mary Andrews and a few other BNFC members were inspired by Percy Fry Kendall, a leading member of the Glacialists' Association and Professor of geology at the University of Leeds, who had done much to promote the study of erratics. They formed a geological committee which was part of the larger field club and between 1893 and 1899 they mapped, named, and collected samples of these exotic blocks of rock, with a view to determining the direction of ice flow in Ulster. Their results were written up by Thompson, as secretary of the geology committee, and published in the *Proceedings of the BNFC* (Thompson, 1901a) and they also contributed their findings to the BAAS committee and reported them to the *Irish Naturalist*, the foremost natural history journal in Ireland at the time (Thompson, 1902, 1906; Wyse Jackson and Wyse Jackson, 1992).

Mary Creese, who has published extensive research on women scientists of this period, has noted that Thompson's major find was a piece of Ailsa Craig microgranite at Moys, inland from Limavady, Co. Derry, which demonstrated the most westerly extent of the Irish Sea Glacier, during the last Ice Age, some 198,000 years ago (Creese and Creese, 2004, p. 51). This distinctive granite, from a small island on Scotland's south-west coast, contains tiny flecks of a blue-coloured mineral, Riebeckite. Small pieces of the rock were carried by the Irish Sea Glacier southwards and deposited with glacial sediments along the east coast of Ireland, and its presence allows geologists to trace the flow of this long-melted glacier.

Today, the Belfast group's collection of erratics is in the Ulster Museum, where the small samples are wrapped in brown paper tied neatly with pink ribbon.

In 1901, Mary Andrews published an important paper on fossil shells and foraminifera (microscopic shells from tiny, single-celled marine animals) collected from glacial sands at Moel Tryfaen in north Wales. The importance of this site is that it is high above sea level and reveals how ice can deposit marine shells at high elevations. Andrews recognised the presence of 12 species, all told, and her findings were utilised by an eminent micro-palaeontologist, Joseph Wright, in a report to the BAAS and reprinted in the *Geological Magazine* (Wright, 1900).

Another British Association initiative was collecting photographs of geological features in Britain and Ireland, and several Irish photographers contributed prints. By the time, the association gathered in Dublin for its 1908 annual meeting, the collection contained 669 images taken in Ireland. The bulk of these were scenes in Co. Antrim (287 images), and the majority of the Irish images had been taken by Robert John Welch, a noted Belfast photographer (Evans and Turner, 1977). Others were supplied by Prof. Grenville Cole of Dublin, Prof. Sidney Reynolds of Bristol, J. St. J. Phillips and W. Grey of Belfast, and by Mary Katherine Andrews. Andrews was an enthusiastic photographer of some ability, and she often assembled sets of Irish views for the annual BAAS meetings. Most of her photographs were taken in Ireland between 1895 and 1907, and they include a quarry at Templepatrick, Co. Antrim, in which a succession of chalk, rhyolite and basalt was exposed; Whitewell Quarry near Belfast where the chalk is cross-cut by a basalt dyke; and the rare orbicular granite at Mullaghderg, Co. Donegal. Andrews also supplied photographs of geological features in Dorset. Many of the photographs of Irish geological features are now in the collections of the Ulster Museum which are listed in an unpublished catalogue. In 1898, W.W. Watts, secretary of the BAAS geological photographic committee, included a drawing from one of Andrews's photographs of a dyke cross-cutting Triassic sandstone in his popular textbook *Geology for Beginners*, which went to several editions.

In addition to their reports on glacial phenomena, the Belfast ladies also published other geological papers. Andrews wrote on dykes in Antrim and Down (1894) in which she published her photograph of the unusual pair of intersecting dykes at Macedon Point near Larne, and on coastal denudation and erosion (Andrews, 1893, 1901b); Thompson published a short note on mussels (1894), considered the nature and identification of White Lias, a Jurassic lithology of Co. Antrim (1900, 1901b), and, in 1910, a biography of her late husband in which is reproduced his pencil drawing of her cutting their wedding cake whilst on a picnic. Appropriately, she is sheltering in the shadow of a large erratic ([Thompson] Christen, 1910).

Igneous studies

Doris Livesley Reynolds (1899–1985) is now remembered by many geologists as the wife of the celebrated geologist Arthur Holmes, who devised the first absolute timescale for the age of the Earth by applying radiometric dating of rocks throughout the geological succession (Lewis, 2000; Wyse Jackson, 2006a). However, Doris was a geologist of considerable ability, which was recognised by her peers when she was awarded the Lyell Medal of the Geological Society of London in 1960. Although not born in Ireland, her parents had moved from Belfast to Manchester shortly before her birth, and she was on the staff of Queen's University, Belfast between 1921 and 1926. It was in the igneous regions of Ulster where she carried out her most important work.

Educated in Essex, Reynolds received her geological education under the redoubtable Catherine Raisin at Bedford College, a constituent college of the University of London that had been established for the education of women (Burek, 2007). Doris gained first-class honours and graduated with a B.A. in 1920 and was immediately employed at QUB by Arthur Dwerryhouse as his research assistant. Soon after, she arrived he moved to Reading, but she remained as assistant to his successor, John Kaye Charlesworth, a glacial geologist. In 1926, she returned to Bedford College as lecturer in petrology and received the degree of D.Sc. in 1927. In 1933, she was appointed by Arthur Holmes to a lecturership in Durham and in 1939, they married shortly after the death of his wife. Unease about this relationship permeated the university and Holmes and Reynolds moved to Edinburgh in 1943 when he was appointed to a Chair in geology (Lewis, 2000). On retirement, they moved to the south of England and Doris died in 1985 having survived her husband by two decades.

Doris published a series of papers on the geology of the north of Ireland. One of her first featured the same Triassic sandstones photographed by Mary Andrews many years earlier (Reynolds, 1928). However, there were her papers on the igneous rocks that brought her most acclaim. Among the first of these was a short article that focussed on the nature and distribution of dykes, which she had mapped throughout the Ards Peninsula (Reynolds, 1931). Moving on to larger field areas she tackled the impressive igneous intrusions around Newry and Slieve Gullion, and in time, she published a number of detailed geological papers and maps of the district (Reynolds, 1934, 1937a, b, 1942, 1943a, 1951). She also contributed to what became known as the 'Granite Controversy' during the 1930s and 1940s, when various geologists debated the nature and formation of granite (Reynolds, 1943b, 1946, 1954). Reynolds also produced a plausible and now accepted account of the mechanism for the formation of rare sodium-rich rocks called albite, which she argued were formed during metamorphism through the introduction of hot fluids (Reynolds, 1942).

Doris Reynolds was a woman with huge energy and, it was said, she was strongly opinionated and feisty. She spent many years working on the geology of her parents' homeland and contributed a large volume of published work on significantly complex geology. Of all the field geologists working in Ireland during the first half of the twentieth century, her contributions were amongst the most significant internationally.

Stars, graptolites, and museum collections

The Dublin Naturalists' Field Club was established in 1886 and over the years, its strengths have largely been in the field of botany. Geology generally received less attention, but between the 1950s and 1970s, two women associated with TCD did much to promote the subject amongst fellow club members. Barbara Miller (née Thompson), who had been a graduate student in geology, served as president of the club in 1965–1966 and led 24 geological excursions between 1957 and 1979, while Veronica Conroy Burns led 19 trips over 27 years until 1986 (Wyse Jackson, 1987).

Veronica Conroy Burns was born at Calvinstown, Co. Kildare on 13 November 1914 and died in the Molyneux Home, Leeson Park, Dublin on 5 January 1998 aged 83 (Wyse Jackson, 1998d, 1998e). Raised by foster parents in Dublin, she lived in a garden flat at 39 Pembroke Road in Ballsbridge, which remained her home for all but the last two years of her life. Largely self-educated, Veronica developed interests in astronomy, classical archaeology, natural history, and geology. She worked for a period as an assistant at Armagh Observatory in the 1940s under its then Director, Eric Lindsay, where among other duties she assembled meteorological records, and she was a founder member in 1938 of the Irish Astronomical Society. Later, she worked in a photographic studio in Dublin. Her wartime Travel Permit Card (now in the collection of the Geological Museum, Trinity College Dublin), issued on 23 October 1944 to allow her to travel between the Republic of Ireland and Britain, describes her as a 'Junior Observatory Assistant'. This she used to travel to work as a laboratory assistant first with Goodman's Industries, Wembley, and later with GEC Research Laboratory, East Lane Wembley.

In 1964, she was appointed to Trinity's Department of Geology, where she was employed by Professor R.G.S. Hudson as a photographic Technician. Later, she assumed responsibility for the Geological Museum and its extensive collection, and immediately redesigned and updated the displays (Figure 19.2). She curated and catalogued some 20,000 specimens and reorganised in particular the mineral collection of which she was especially fond. She was ideally suited to museum work being neat, meticulous, and interested in visitors. She was never happier than when showing children around the museum. Years later, visitors continued to ask for her, and for many who graduated to employment as professional geologists Veronica and her museum was their first experience of geology.

Veronica Burns's main geological skill was as a fossil collector, and in this she excelled. On Saturday mornings, she would take a bus to the collecting localities and spend all day splitting rocks in her search for good specimens, stopping only briefly to eat a sandwich. In the early 1960s, she began to collect Silurian graptolites (fossils of colonial marine animals), from the coastal area between Balbriggan and Skerries in north Co. Dublin and from small quarries and exposures inland. These rocks generally contain few fossils, but through her tenacity and patience, she was able to assemble a large collection of about 2,000 specimens. Through the influence of Hudson, she also focussed on Feltrim Hill, still in north

Figure 19.2 Veronica Burns (Geological Museum, Trinity College Dublin).

Co. Dublin, where she collected a large number of Lower Carboniferous fossils from the Waulsortian carbonate mudmound and its intercalated shales. These were used by Hudson, Michael Clarke and George Sevastopulo (1966) in their paper published in the *Scientific Proceedings of the Royal Dublin Society*.

Burns kept an album of photographs that she had taken of the quarry at the time, and it contains two fascinating images: one is of the windmill that once stood on top of Feltrim Hill, perched precariously on an isolated tower of limestone around which quarrying had taken place, and the second is of Veronica herself brandishing her hammer at a mechanical digger, and captioned ‘The Rivals’. The album is now in the Geological Museum in Trinity, along with her fossil collection.

Veronica Burns was modest about her achievements and only when Barrie Rickards, a graptolite expert, was appointed as a geology lecturer in Trinity College Dublin in 1967, was the importance of her graptolite collection recognised. With Rickards and Jean Archer in 1973 and with Rickards in 1993, Burns published two papers in which her collections were used to determine precisely the age of the rocks around Balbriggan that are 440–420 million years old. In the latter publication, a new graptolite subspecies, *Monograptus flemingii* (Salter) *warreni* Burns and Rickards, 1993, was described and illustrated.

Burns was deservedly elected an honorary member of the Irish Geological Association and an honorary life member of the Dublin Naturalists' Field Club. Although increasingly incapacitated in the last years of her life, Veronica remained cheerful and interested in the activities of her family and friends. She had a wide circle of friends and colleagues and her enthusiasm for geology and astronomy never wavered.

GEOLOGICAL COLLECTIONS

20

GEOLOGICAL MUSEUMS AND THEIR COLLECTIONS: RICH SOURCES FOR HISTORIANS OF GEOLOGY

Annals of Science 56(1999), 417–431.

Geological museums and collections: development of a resource

The development of geological museums in Great Britain has been demonstrated to consist of four periods of contrasting fortunes (Knell, 1996). Knell's periodisation offers a starting point for an effort to gain perspective on the uses of museums and collections for investigators in the history of geology. For various reasons, museums and collections have tended to be underutilised in such historical research. Drawing on examples from British and Irish geology, I shall illustrate some of the ways the work of historians of geology can be enriched through materials in geological collections.

The earliest geological collections date from the sixteenth century when collectors were primarily concerned with acquiring mineralogical specimens (Torrens, 1985; Wilson, 1994). The largest collections, and thus often the most significant, were developed by members of the numerous royal houses across Europe, and by members of the aristocracy or the diplomatic corps. While many individuals purchased their specimens from dealers, others travelled widely in order to collect fine geological specimens for their cabinets. These were largely held in domestic museums.

Public museums on the scale with which we are now familiar emerged in the late 1700s, which marked the beginning of the golden age of collecting and which continued largely unabated into the late decades of the nineteenth century. From the 1820s onwards, in Britain, geology had become an important subject for study on account of the development of industry and thus the necessity to find raw materials and fuels. New and wondrous fossils were being discovered and described, and the basis of stratigraphical geology was being laid down by several British geologists. The first rise of importance for geological museums began, according to Knell (1996), during this 'heroic age' of geology, at a time when geologists

DOI: 10.4324/9781003475552-25

and their finds created 'perpetual excitement' amongst the press and hence the general public. Throughout Britain and Ireland, and no doubt elsewhere, many public-spirited city-based philosophical, literary, and scientific institutions and societies emerged, and many of these established geological museums for the benefit of their members or for the public at large (Morrell, 1994).

These institutions, often concerned with technological and agricultural development as well as with public education, acquired geological collections. The holdings of these foundations, which included the Dublin Society (1731), the Bristol Institution (1809), the Yorkshire Philosophical Society (1822), and the Cheltenham Literary and Philosophical Institution (1833) (Torrens and Taylor, 1990) to name but four, formed the basis of many significantly important geological collections still available for study. At the same time, specialist geological societies emerged after the lead given by the Geological Society of London in 1807. It was followed by the Royal Geological Society of Cornwall (1814), the Geological Society of Dublin (1831), and Geological Society of Edinburgh (1834) and others in provincial Britain. However, by the 1850s, geological studies had become more specialised and geology fell from public attention and favour. Many societies which had formed on the crest of geological popularity three decades earlier were now in severe financial difficulties and their collections were reduced in size (such as those of the Tunbridge Wells Literary and Scientific Society) (Gill and Knell, 1988) or taken over by other institutions; those of the Geological Society of Dublin were given to Trinity College Dublin in 1848 (Herries Davies, 1965; Wyse Jackson, 1992b). As a result, geological museums in general declined in importance at this time (Knell, 1996, pp. 34–35).

However, the trough into which geology had descended was upturned a decade later as Victorians in Britain and Ireland were gripped by the fascination for all aspects of natural history (Allen, 1976; Barber, 1980). Fern and fossil collecting amongst other hobbies became popular with the masses and natural history museums (which included geology as an essential component) and zoological gardens became regular places to visit. A contemporary survey of 168 rate-supported, subscription-supported, government-supported, and university museums shows that the majority contained geological specimens (Greenwood, 1888). This listing is reproduced in Greenwood (1982). The museum in Birmingham claimed to attract 18,000 visitors a week. In Britain and Ireland, this upsurge in popularity for natural history resulted in the genesis of the field club movement, which in turn helped sustain it. In Ireland, at least five field clubs were active, two of which—Dublin and Belfast—are still extant. However, geological museums went through another decline after the end of the First World War when geological collections were either seriously neglected or all too often simply disposed of.

Within the British and Irish universities, geology as an academic subject was formalised in the mid-nineteenth century and teaching of geology was often through the medium of the museum and its holdings. Many universities assembled major collections: Oxford, Cambridge, King's College, London, and

Trinity College, Dublin all had important museums. Typically the specimens students were expected to recognise were displayed in regimented rows of polished oak or mahogany glass-topped cabinets. In the 1950s considerable philosophical changes in the methods of instruction in geology saw a shift from museum-based studies to laboratory and field-based studies, where students handled specimens for themselves, although such methods had been advocated earlier by Thomas Huxley (1825–1895) and implemented earlier elsewhere by teachers such as Baron Cuvier (1769–1832) in Paris, and Louis Agassiz (1807–1873) at Harvard. This resulted in a decline in the use of university museums, some of which came to be viewed as expensive white elephants. In Trinity College, Dublin the fine museum gallery designed by the noted Ruskinian architect Benjamin Woodward was converted in 1954 into laboratories (Wyse Jackson, 2019), and many, but fortunately not all, specimens collected by many pioneers of nineteenth-century geology such as Richard Griffith (1784–1878), Thomas Oldham (1816–1878), and John Phillips (1800–1874) were discarded (Wyse Jackson, 1992b). While some downgrading of university collections did occur from the 1950s onwards, many collections were saved, either through the considerable efforts of a few individuals or, ironically, as a result of sheer neglect and indifference. In the 1980s in Britain, geology at university levels underwent radical review, and some departments were been shut down on the recommendations of the Oxburgh Report. Nevertheless, in the face of such difficulties, many collections were relatively safe, and the future for some selected university collections (those at Birmingham, Cambridge, Glasgow, Manchester and Oxford) became more secure with funding provided in the early 1990s by the Universities Funding Council for storage and curatorial cover; although at the time, the choice of selected museums drew some questions (Taylor, 1990a, b).

At the present time, knowledge of the holdings and needs of geological museums is greater than at any time in the past. In the UK, public awareness and appreciation of geological museums continue to rise, and much of this has to do with the efforts made by the museum profession itself and with the establishment by concerned curators and geologists of the Geological Curators' Group in 1974. In 1981, in his landmark survey of 569 museums in the UK, Philip Doughty conservatively estimated that 283 of these housed between 25 and 30 million geological specimens, and that the Geological Survey collections and those in the Natural History Museum, London were the most numerous with approximately 10 million specimens each. At that time, only 16% of the museums employed full-time professional staff (Doughty, 1981). In the 1990s, further surveys compiled additional comprehensive information pertaining to geological holdings for 100 public or university museums in Britain and Ireland, and it was found that less than half of these institutions employed a specialist geological curator (Nudds, 1994; Parkes and Wyse Jackson, 1998). Clearly, many collections are without specialist care, and it is these collections, termed 'orphan collections' (Collins *et al.*, 1996) which may be at greatest risk of being broken up, dispersed, or disposed of during times of financial tightening in the museum and university sector.

Geological collections and geological historians: use or non-use?

Are geological collections used by historians of geology as sources of information? Naturally, there are examples of when they have been utilised, and others of when they have not. To attempt a quick answer to the question, I rapidly scanned the listings of sources given in several reputable volumes on the history and philosophy of geology published between 1997 and 1992 (Porter, 1977; Greene, 1982; Rudwick, 1985; Secord, 1986; Laudan, 1987; Dean, 1992). In all the books, consulted listed manuscript or printed sources, no authors documented which geological collections and specimens, if any, they had examined. Detailed examination of the texts may, however, reveal the use of specimen-derived data. Without presuming to question the value of these studies, or to undermine the scholarship of the authors involved, I do wonder whether a systematic effort to utilise collections and specimens might have contributed significantly to the strengthening and improvement of these excellent studies. While this is difficult to determine in every individual case, it is the conjecture of this geological curator that such use of collections may add to the overall picture, and might influence the conclusions reached. Indeed, one need not claim a comprehensive knowledge of the literature in the history of geology to observe, as I believe few would deny, that historians of geology have tended overwhelmingly to rely on published works papers and books, as well as maps—and not uncommonly also on manuscripts, as their primary source material. It is my contention, of course, that geological specimens, although often overlooked, are primary sources of valuable information too. D.T. Moore (1982) in an introduction to his catalogue of rock collections in the Natural History Museum acquired before 1918 acknowledges the historical importance of rock specimens. But how often has his valuable paper been trawled by geological historians in the search for specimens? Might we not expect researchers to utilise the full range of geological resources that remain, in order to produce a comprehensive treatment and synthesis of the subjects under examination?

Certainly, there have been historical papers that make reasonable use of geological specimens in their development—but the percentage of such publications is small when measured against the total published works within the sphere of the history of geology. An excellent example of historical research that makes use of a collection of mineralogical specimens—in this instance to confirm an eighteenth-century exchange of geological specimens between the Italian scientist Giovanni Arduino (1714–1795) and the German Nathanael Gottfried Leske (1751–1786) (Vaccari and Monaghan, 1993). Similarly, the rediscovery of an important collection of Jurassic and Cretaceous fossils was the stimulus for publication of a valuable paper about its owner Etheldred Benett (1776–1845), her scientific correspondents, and her own publications (Spamer *et al.*, 1989).

These honourable instances notwithstanding, the overwhelming majority of papers and books on geology's history make little or no reference to actual geological collections. Why should this be? How can historians of geology truly understand the significance made by their subject to the development and understanding of the science, if they access their published record only and do not examine all available sources of data? Why has there been a tendency to ignore what historical information geological specimens can yield? Perhaps, it is due to a number of factors of which I highlight three.

State of geological collections

It is undeniable, unfortunate, that many geological museums worldwide do not have the required resources to house their collections in a manner to their liking; commonly curatorial staffs are thin, and some museums have no curators at all. Since 1974, the Geological Curators' Group has fostered a greater awareness of geological collections and still acts as a forum and focus for those in the profession, and this group has now attracted a worldwide membership. The Group has been emulated by other organisations, such as the Society for the Preservation of Natural History Collections (SPNHC) of North America, with similar concerns for geological collections; these bodies have done much in the past five decades to highlight the plight of collections, have promoted the revival of good curatorial methods and collection documentation, and have promoted research on the history of collections and collectors. In some cases neglect, often benign, is not actually a bad thing: certain collections are still in their original state and have not been recatalogued, relabelled, or rehoused processes which often lead to serious loss of historical data.

Examination of the journal *The Geological Curator*, published by the GCG, attests to the efforts made to make geological specimens more relevant in historical studies: frequently important specimens thought to be lost are tracked down and their significance highlighted.

Other important initiatives include the publication of directories of collections. Sherborn's (1940) volume was the pioneer; more recently, *World Palaeontological Collections* (Cleevely, 1983) and *World Directory of Mineral Collections* (Zwaan *et al.*, 1974, revised 1977) are the main standards. These are essential (but unfortunately rapidly dating) sources for specimen information, as are the many catalogues of museum specimens, especially of type and other status material (Bassett, 1975; Deisler and Bassett, 1997). Considerable data on collections in the UK were assembled during the Federation for Natural Science Collection Research (FENSCORE) project of the 1970s and 1980s, and this made available a number of valuable reports which together constitute a rich source of information for historians of geology (Hancock and Pettitt, 1981; Davis and Brewer, 1986; Hartley *et al.*, 1987; Stace *et al.*, 1988; Bateman *et al.*, 1993; Walley, 1993). Additionally, volumes such as the *Directory of British Geological Museums* (Nudds, 1994) are useful starting points for tracking specimens. Collections in Britain and Ireland are now well documented. The methods and lessons learned from these initiatives need to be taken on board by the geological and historical community in other countries so that more data pertaining to the state and status of geological collections worldwide are forthcoming. Given the digitisation capabilities offered by the internet, there is a great opportunity to assemble all these geological collection data into one searchable online resource.

Geologists vs historians

The study of the history and philosophy of geology is undertaken by both historians and geologists, who by the nature and background of their training may approach the subject in often different ways. There are of course notable instances of scholars with strong formal backgrounds, or extensive practical experience, in both history and the earth sciences; but these are comparatively rare cases. Geologists are trained to be objective and to infer conclusions from the study of specimens and landscapes, whereas historians tend to be more analytical in their approach. Geologists are not trained to ingest copious amounts of textual material, which often requires deep and long consideration in order to arrive at its meaning.

Historians tend to be more familiar working with paper sources—printed as well as manuscript—and may reach a greater understanding of and from these sources than can many geologists. It is therefore natural that historians should be more reliant for source information on printed and archival materials as their primary sources of information. On the negative side historians often lack the expertise necessary to analyse available geological specimens. It may be observed, however, that in some circumstances, the historian's technical ignorance can actually be an asset: unlike their geological counterparts, historians without scientific training may sometimes more easily escape the influence of modern geological ideas and thus more readily approach the mindsets of early geologists.

Modern geological papers frequently present the results of experimental or field results which to geologists make easy reading. But modern scientific training may actually contribute to some difficulty in understanding papers authored by the geological pioneers of the eighteenth and nineteenth centuries, which are frequently long, verbose and based on fewer observational and experimental data than their modern equivalents or on data reported in a fashion puzzling to those accustomed to current standards of scientific practice. Again, the historian may sometimes have an advantage through an accustomed adaptability to past conceptual and verbal habits. It is important, however, that both historians and geologists work together on problems and projects in the history of geology. Such collaboration will allow for fuller and more complete analyses and can be expected to yield conclusions more incisive than if this work was undertaken by one or the other group in isolation.

In recent years, historians and geologists interested in the history of geology have developed closer working ties through the existence of a number of international and national bodies. The International Commission on the History of Geological Sciences (INHIGEO) acts as a global umbrella organisation, and national bodies such as the History of Geology Group (HOGG) of the Geological Society in the UK and the History of Geology Division of the Geological Society of America, and specialist societies including the Society for the History of Natural History, all promote study into the history of geology, which has led, amongst some aficionados, to re-examination of geological museum holdings.

Attitudes of modern geologists to the work of their predecessors

Many geologists carrying out research today cannot accept that research carried out in the past in their specialist fields has any significance for them. In their view, such work by their pioneers is simply out-dated, outmoded, and old-fashioned. If interpretation of specimens from the past is usually beyond the competence of historians, many geologists in possession of the necessary expertise frequently see little point in the exercise. This 'cultural barbarism'—an expression coined by Peter Medawar in 1979—is arrogant and illogical. A fundamental tenet in science is the need for viable checking and reproducibility of results. Reanalyses may not be undertaken for some time after the original research but require preservation of the original material worked on in order to be of any value.

Palaeontologists are all too aware of the importance of preserving type material—this allows confirmation of the original diagnoses of fossil species erected in the past. For instance, it is not unusual for workers on Carboniferous faunas to consult the specimens described by Frederick M'Coy (1823–1899) in his monographs of 1844 and 1846, specimens housed in the National Museum of Ireland (Wyse Jackson and Monaghan, 1994).

Similarly, stratigraphical conundrums that appear in the literature can often be answered by means of examination of the original fossils collected by the original workers. Incorrect original taxonomic determinations often lay at the root of problems which frequently disappeared in the light of new determinations. *Trinucleus*, a small fringed trilobite, is thought to be diagnostic of Ordovician strata. However, in Ireland, several records of the trilobite from Silurian sediments were published at the turn of the twentieth century. Recently, some of the specimens were traced and shown to have been misidentified, being either other trilobite taxa or rugose corals, or being contained in reworked or transported blocks of older sediments (Parkes and Owen, 1994).

At the beginning of the present century, measurement of the radioactivity of various rocks was carried out by several scientists, including R.J. Strutt and John Joly (1857–1933). Joly measured the volume of radium in a large suite of rocks including specimens from the Indian Deccan plateau, the volcanoes of Italy and the granites of Ireland and Britain, and showed that the quantity of radium was greatest in recently erupted lava (Joly, 1909c). He recognised that radioactivity might hold the key to determining the age of the Earth, but it was the English geologist Arthur Holmes (1890–1965) who, in 1911, produced the timescale that we are now familiar with (Lewis, 2000). Is Joly's work worth re-examination? How accurate were Joly's quantitative methods? With modern analytical equipment would similar results be forthcoming from his material? Unfortunately, it would prove impossible to recollect from some of his sources, so redeterminations would have to be carried out on Joly's original material. Considerable insights into the foundations of geology, and the background to geological theory, may be gathered by historians of geology and by geologists researching today through detailed examination of the printed, manuscript, and object-based contributions of former geologists. Cultural barbarism has no place in the modern research library or laboratory research has only progressed to the stage reached today through the endeavours of our research ancestors and their legacy is valuable and should be enjoyed, utilised, and understood by contemporary historians and geologists.

The value of geological collections for the study of the history of geology

The close study of minerals, rocks, fossils, and other geological materials can reveal much about their history and their significance to past geological research. They allow for testing and reproducibility of original results and observations, for comparison of modern and older classification schemes. Most significantly, they can suggest to modern researchers the reasons why certain ideas originated in the minds of the early geologists.

The perceived and accepted significance of collections may wax and wane with time. Some collections have suffered damage or neglect because they were deemed at the time to be of low significance and thus of low priority when it came to allocating time to their curation and adequate facilities for their storage. All geological specimens were acquired for particular reasons, but perhaps due to a combination of poor collecting or curatorial practices, much significant information pertaining to the collection may have been either not originally available or subsequently destroyed. This radically decreases the value of the collection. However, possible rehabilitation of collections is not impossible, although this may entail laborious and time-consuming research by museum curators or geological historians.

Collections which can be of greatest use to historians of geology are those with good documentation. Thus original attached specimen labels, tray labels, original field sheets and notebooks, manuscript catalogues, and even original packing materials can all be vitally important. Equally, newer labels and other associated items can be important in tracing the history of collections. An undocumented and unidentified piece of granite is simply that: perhaps attractive but worthless in terms of scientific history.

Research on collections in geological museums should concentrate in two areas: first, the extant collections, if in a disorganised state, should be rehabilitated. Specimens should be cleaned, identified, reboxed, and catalogued, and all available documentation (old labels, tray cards, catalogues etc.) sorted, matched with specimens, and conserved. Second, if material has been discarded in the past, then an attempt should be made to determine precisely what this was. Examination of manuscripts and printed records might allow for this to be done successfully.

Once rehabilitation of specimens and/or collections is complete, then it is more probable that their significance to the development of geological thought will be recognised.

Below are some case studies where the study of extant specimens or documentation relating to lost geological collections has led to a greater understanding of the collector's geological research and output, and also to an awareness of some cultural and social factors that marked their history.

Case study 1: George Graydon and tracing lost collections

Frequently large portions of collections are lost, and it is often difficult to work out their original composition. Nevertheless, it is a worthwhile exercise attempting to ascertain how collections were dispersed, as this may give reasons for their loss and may also reveal unknown links between institutions and individuals.

In 1790 and 1791, the Dublin cleric George Graydon (c. 1753–1803) travelled to Italy to collect volcanic specimens from Vesuvius, and more unusually from the region of extinct volcanoes in the north-east of the country (Vaccari and Wyse Jackson, 1995; Chapter 3 herein). He did this to allow his colleagues in the Royal Irish Academy to compare the Italian specimens with material from north-east Ireland around which much debate centred concerning the origins of some igneous rocks. Graydon returned with over 3,500 specimens, but today only nine have been traced. Fortunately, his catalogues exist in Trinity College Dublin, allowing researchers to determine the composition of his collection. These give some indication of his thoughts on the nature of basalt in north-east Ireland.

In 1794, Graydon presented the choicest specimens (about half the total number) to the Royal Irish Academy for their museum and retained the rest for himself. What effect did these collections have on the members of the Academy? Little, it would appear. Some members, such as the Reverend William Hamilton accepted that the basalts were the products of volcanoes and other igneous phenomena. Others, such as Richard Kirwan and the Reverend William Richardson refuted the volcanic origin of basalt (Wyse Jackson, 2000; Chapter 2 herein). While Kirwan praised Graydon's efforts in acquiring the specimens, it is surprising that in his and Richardson's papers which deal with igneous phenomena, and were published by the Royal Irish Academy, no mention is made of Graydon's collection (Kirwan, 1793; Richardson, 1803a, b, 1806a, b). Graydon had seen that the recent lavas that he had himself collected from the slopes of Vesuvius bore a close similarity to the ancient Antrim basalts, and while he did not voice an opinion one must assume that he considered that the latter were similarly the products of volcanic activity, rather than precipitates from water.

In 1801, Graydon's specimens were transferred from the Royal Irish Academy to the Dublin Society, which now forms the bulk of those in the National Museum of Ireland, where six fossil fish collected by Graydon from the Eocene of Bolca, north-east Italy, are still extant (Vaccari and Wyse Jackson, 1995).

Of the half of his collection which Graydon retained, some he gave away to personal contacts, while the remainder was sold to Trinity College by his widow Elizabeth soon after his death. At the same time, a small suite went to the Royal Dublin Society. The Swiss geologist J.F. Berger (1779–1833) who worked in north-east Ireland (MacArthur, 1990), and who published some of Graydon's findings on Monte Somma in 1816, was given eleven specimens from Trinity College in about 1811, and these made their way to the Geological Society of London and later to the Natural History Museum in London where only one specimen remains. The whereabouts of the remaining specimens formerly in Trinity College Dublin are unknown (Wyse Jackson and Vaccari, 1997).

Study of the fate of missing collections can reveal unexpected information concerning the priorities of museums, the attitudes of users, and contemporary perceptions of value.

Case study 2: Robert Jameson and understanding early geological terminology

Correct understanding of early geological printed works and manuscripts can be difficult to achieve when lithological and mineralogical terminology unfamiliar to the reader is used. This is a prevalent problem in early texts when mineralogical and lithological names had not become standardised; in many cases, names were descriptive in terms of the mineral habit and constituents (e.g. ruin agate, roe stone, or copper foam). For historical purposes, the compilation by A.H. Chester of mineral names is important, as is *Hey's Index of Mineral Species* (Chester, 1896; Clark, 1993). Arkell and Tomkeieff's *English Rock Terms* (1953), and Tomkeieff's later *Dictionary of Petrology* (1983), are equally valuable, but the historical usages these works treat are principally of vernacular terms, not based on prevalent scientific knowledge. While nevertheless valuable, these pioneering works were largely bibliographic and etymological in nature, and it is suspected that the authors did not always examine actual material to which these early names had been given. Berg-Madsen (1989) discussed the problem of terminology in relation to the dark bituminous limestones of Scandinavia which have been given synonymous names including stinkstone, anthraconite, and orsten. While orsten is still used for these limestones in northern Europe, she showed that modern interpretations of the terms deviate markedly from the original meaning. Throughout early texts, terms are used which may not be known to latter-day geologists or historians of geology, or else terms are used today which are at variance with past usage. For historians of geology, this is a perpetual problem, because, without complete understanding of geological terminology, past and present, it is impossible fully to understand geological theory. One solution to this problem would be for researchers to examine collections of minerals and rocks assembled by authors of early geological texts, determine the old published names, and correlate these with the modern mineralogical and petrological names applicable to the specimens.

Robert Jameson (1774–1854), Professor of Natural History at the University of Edinburgh from 1804, studied mineralogy under Werner at Freiberg. On his return to Scotland, he published many volumes on mineralogy including the three-volume work *System of Mineralogy*, which went to three editions. In 1821, Jameson's *Manual of Mineralogy* appeared, a book which is of interest here on two counts. First, it gives the original German names coined by Werner, and their English equivalents. Second, it contains a section in which Jameson gives a description of mountain rocks which are arranged to allow the reader a geognostical or structural and compositional view of the Earth. It is accompanied by a listing of 120 rock specimens from the Hartz Mountains of central Germany (Figure 20.1A), which are

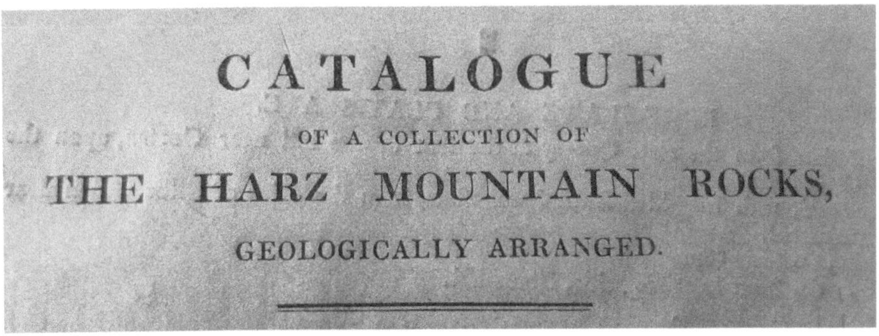
CATALOGUE
OF A COLLECTION OF
THE HARZ MOUNTAIN ROCKS,
GEOLOGICALLY ARRANGED.

Figures 20.1 (A) Title of Hartz Mountain specimens (1825)

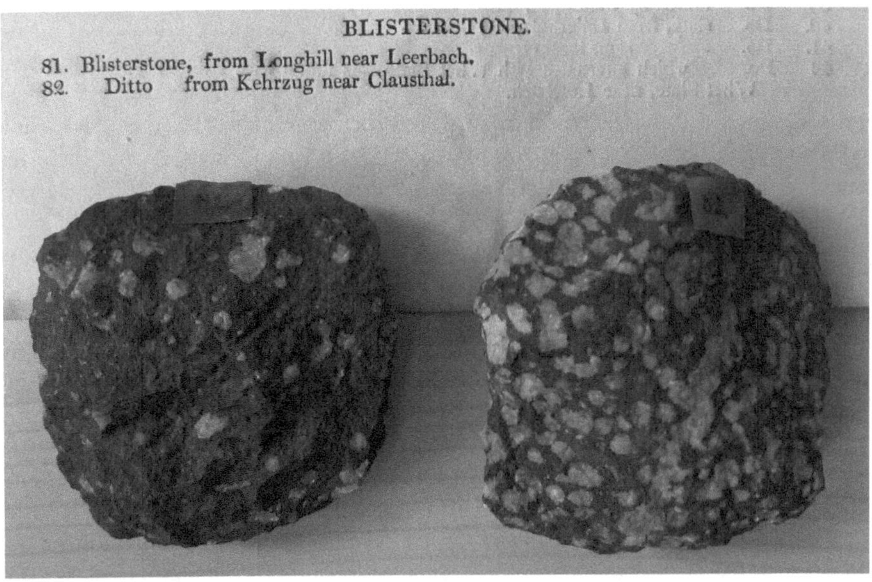

Figures 20.1 (B) Two specimens of 'Blisterstone' from the Hartz Mountains, Germany (Geological Museum, Trinity College Dublin).

catalogued under three categories: (1) primitive rocks, which include granite, gneiss, horn-rock, and quartz-rock; (2) transition rocks, which include limestone, ironshot, greywacke, jasper and blisterstone; and (3) floetz rocks, including conglomerates, marls, and muschelkalk. While most of these lithologies are familiar, some such as blisterstone may prove to be new terms to most geologists and historians, given that it does not appear in a number of modern dictionaries or glossaries of geology. What is blisterstone in terms of modern nomenclature? In the Geological Museum, Trinity College Dublin is stored a set of these Hartz Mountain rocks, and close examination of specimen numbers 81 and 82: blisterstone—reveals them to be amygdaloidal basalts (Figure 20.1B). The overriding value of collections such as the Trinity College suite is that they illustrate lithological classifications formerly used and they allow for correlation and understanding of terminologies used in the past and at the present time.

Case study 3: Social interaction between collectors in Dublin in the 1800s

Prosopography is the study of collective characteristics of people by way of studying their lives; this approach has been attempted in a limited fashion in the history of science (Shapin and Thackray, 1974; Pyenson, 1977; Morrell and Thackray, 1981; Livesey, 1998). Allen (1990) has argued that in applying the technique to the study of the history of natural history, much hitherto hidden can be gathered. Rudwick (1988) on members of the Geological Society of London, Ciancio (1995a) on the relationships between British and Italian naturalists, and Creese and Creese (1994) on women geologists in the nineteenth century, can be said to have used this prosopographical method. As time passes, it becomes increasingly difficult to detect or decipher cultural and social patterns. Looser geological prosopographical studies include many scientific biographies (Herries Davies and Mollan, 1980; Rupke, 1994; Desmond, 1994, 1997) and institutional histories such as those of the Geological Society of London (Woodward, 1908; Herries Davies, 2007; Lewis and Knell, 2009). Others include investigations of past controversies particularly on the nature of various rock types such as the Old and New Red Sandstone (Diemer, 1996) and those in northern Scotland (Oldroyd, 1990) or the establishment of geological periods (Rudwick, 1985; Secord, 1986). All these studies contain valuable social and scientific insights. While prosopography strictly refers to the study of people, a similar method can be applied to geological collections and collectors, and occasionally these cultural and social patterns become evident through their examination. Exchanges between Victorian naturalists and museums were quite frequent and worthy of investigation. Harper and Parkes (1996) document an example of specimen exchange in Ireland.

Richard Kirwan was instrumental in the Dublin Society's acquisition of the Leskean mineral collection in 1792. This collection, which numbered several thousand specimens, was assembled by Nathanael Leske between 1782 and 1786 (Vaccari and Monaghan, 1993) The presence of the Leskean collection in Dublin fostered interest in the science; many people and institutions, including Trinity College, began to amass mineral collections. One such collector was the Honourable George Knox (1765–1827), a member of parliament for the University of Dublin (Wyse Jackson, 2006b; Chapter 21 herein). In 1827, his alma mater paid £500 to purchase his extensive collection of specimens from Europe, Greenland, and North America. Examination of the specimens and accompanying hand-written catalogues shows that Knox arranged his collection using a classification similar to that of Werner and provides answers to some questions about the way it was assembled. How did Knox acquire the specimens from Greenland? Could he have received them from the Greenland explorer Sir Charles Lewis Giesecke (1761–1833) who had secured appointment in 1813 as Professor of Mineralogy to the Dublin Society? Study of all the extant Greenland material in Trinity College has shown that it belonged previously to two collectors—to Knox and to another. The mineral species present included Cryolite and Sappharine, unknown in Europe before their discovery by Giesecke. Confirmation that Knox had been given the material by Giesecke came in an entry in his hand-written catalogue. Could the rest of the Greenland specimens in Trinity have come from Giesecke himself? It is known that he donated a great deal of material to museums across Europe, so why not to Trinity? Examination of a hand-written list of specimens, in Giesecke's hand, confirmed that the source of the second suite of Greenland specimens was indeed Giesecke (Wyse Jackson, 1996b; Chapter 22 herein).

The Geological Society of Dublin was established in 1831 and was granted a royal charter in 1864 from whence it was styled the Royal Geological Society of Ireland ([Herries] Davies, 1965). For nearly 60 years, it served as an important forum for geological debate. In order to familiarise its members with geological material, it began to assemble a museum soon after its inception, and soon it contained rocks from north-east Ireland, as well as other regions, including the volcanic districts of the Rhine. Donors to the collections included John Phillips and the Earl of Enniskillen (1807–1886). The museum contents were held from 1848 in the collections of Trinity College Dublin. The passage of generations has obscured the identity and provenance of some of these items on display in Trinity College, yielding puzzles which can, it turns out, be solved through research. Examples include some wall-mounted marine reptile specimens whose identity was long unknown; a recent close examination showed them to be copies of a plesiosaur scapula and sternum from England first described by Thomas Hawkins (1810–1889). Why are they in Dublin? Subsequent perusal of the *Journal of the Geological Society of Dublin* revealed that Hawkins offered the replicas as a gift. Why? Could it have been for reasons of self-aggrandisement certainly a possibility? Could Hawkins have had contacts in Dublin? Yes, apparently so. He was friendly with Robert Hutton (1784–1870), a Dublin merchant and Member of Parliament, who had presented the society a year earlier with a replica of a complete plesiosaur skeleton—one originally illustrated by Hawkins (Wyse Jackson, 2004).

The use of archival and specimen data helped sort out the provenance of the Greenland specimens, while tedious taxonomic identification allowed the puzzle of the provenance of the plesiosaur bones to be untangled and solved. Hidden cultural and social interactions and bonds, as have been illustrated above, can be revealed through the unfashionable study of geological specimens.

Case study 4: Geological travellers

On account of the nature of the subject, it has always been necessary for geologists to travel widely (see Wyse Jackson, 2007a for an assessment of many geological expeditions). While some geologists are known to have made few expeditions, others such as Sir Charles Lyell (1797–1875) made long trips to exotic locations including North America (Dott, 1996; Wilson, 1988), Murchison to Russia (Collie and Deimer, 2004) trips which moulded their thinking on geological matters. Others such as Ferdinand Hayden in the United States conducted fieldwork as part of official Government surveys (Deibert and Breithaupt, 2016) The American geomorphologist William Morris Davis (1850–1934) spent a great deal of time examining the landscape viewed from trains traversing the North American continent. Others walked, bicycled (Wyse Jackson, 2007b. Chapter 18 herein), took carriages, paddled canoes (Spalding, 2007), or sat astride camels (Taquet, 2007).

One of the major sources of information on geologists' itineraries and the geological features they encountered are the notebooks, diaries, letterbooks, commonplace books, and sketchbooks kept by these travellers which often record additional insights into antiquities, customs, and even everyday life. Draft and printed maps are rich sources of data about geologists' travels, as are photographic collections. From the geological specimens collected and housed in museums, it is possible, even if admittedly difficult, but rewarding, to reconstruct the itineraries of geological expeditions and individuals and to help understand and interpret any scientific writings that emerged from these expeditions.

As has been eloquently shown by David Oldroyd (1990), it is necessary to revisit the sites examined by geologists in the past if an accurate appreciation of their work is to be gained. Confining ourselves to the reading of published papers provides an incomplete view; examination of particular exposures or landscapes can illuminate why certain conclusions were drawn from them. Field notebooks and other manuscript sources are important sources of information for reconstructing fieldwork schedules but so too are geological collections.

Conclusion

Interest in the history of geology has begun to influence the way geological collections are perceived, although their value is still underrated by many. The work of groups such as the Geological Curators' Group has played an important role in fostering good museum management and collection documentation, which should have beneficial spin-off effects for historians of geology.

Geological collections are primary source materials and can give valuable information about geological localities, the collectors of material, relationships between collectors, institutions, museums and places of learning, and cultural and social factors. Study of geological material can also reveal interesting details relating to the development of geological methods, such as thin section making, which are still important in everyday geological investigation.

Most important of all, examination of geological specimens held in museums in tandem with printed sources may significantly enhance our understanding of the complexities of the development of geological thought.

21

THE HONOURABLE GEORGE KNOX (1765–1827), PARLIAMENTARIAN AND MINERAL COLLECTOR: HIS COLLECTION IN TRINITY COLLEGE, DUBLIN, IRELAND

The Mineralogical Record 37(6) (2006), 543–551.

Introduction

The University of Dublin was founded in 1592 with one constituent college: Trinity College. The Dublin University Museum was founded in 1777 to house Polynesian artefacts collected during the voyages of Captain James Cook (Ball, 1853; McDowell and Webb, 1982). The driving force behind its establishment is thought to have been Rev. William Hamilton (1755–1797), one-time Fellow of the University, who is perhaps better known for his geological observations of the Giant's Causeway, and his arguments that these basalts are of igneous origin (Hamilton, 1786).

The Geological Museum in its present form dates from 1857, when the Dublin University Museum was subdivided into a number of smaller units. For nearly a century, the Geological Museum occupied a large room in the building constructed for the University Museum at Trinity College. In 1953, this large room was subdivided and converted into geological laboratories, and the collections were moved to a smaller room nearby, where the Geological Museum is presently situated (Gill, 1956).

The Geological Museum is best regarded as a research and teaching museum, and its collections are displayed for the benefit of university students, schoolchildren, and the general public. Material from the Museum is utilised in undergraduate teaching programmes and is frequently borrowed for academic research. A comprehensive account of the history of the Geological Museum and its holdings is given in Wyse Jackson (1992b).

DOI: 10.4324/9781003475552-26

This article briefly outlines the historical development of the Geological Museum and the origins, scope, and extent of its collections and focuses on the eighteenth-century mineralogical collection of the Honourable George Knox.

Acquisition of the Trinity College collections

Within a decade of its founding, the Museum had acquired some geological material, notably fish fossils from Monte Bolca, Italy, collected by a Dublin cleric, George Graydon (1753–1803) in 1794 (Graydon, 1794; Vaccari and Wyse Jackson, 1995 [Chapter 3 herein]). By 1807, the Museum housed a collection of 1,089 minerals, some petrological specimens, including rocks from Vesuvius also gathered by Graydon, and some fossils (Stokes, 1807).

For nearly 70 years following 1780, the Museum had no permanent curator, although Whitley Stokes, a lecturer in natural history whose political activity brought him into conflict with the University authorities, was responsible for the collections from 1816 until 1844. Few important collections were added to the Museum during this period, with the exception of the large Knox mineral collection, and some Silurian and Jurassic fossils presented by John Phillips (1800–1874), who served in 1844–1845 as the first Professor of Geology at the University.

The welfare of the Museum improved through the efforts of Robert Ball (1802–1857), a noted naturalist who in April 1844 was appointed the Museum's first full-time director; at the same time, responsibility for the geological and mineralogical collections was entrusted to the incumbent professors of Geology and Mineralogy, respectively.

Ball encouraged many geologists and naturalists (including the Earl of Enniskillen, Thomas Oldham, Sir Richard Griffith, Joseph Bette Jukes, Baron Cuvier, and J.D. Hooker) to donate geological material (Ball, 1846, 1847, 1848, 1853). He also arranged for the entire museum of the Peripatetic Geological Society of Dublin (later the Royal Geological Society of Ireland) to be housed at Trinity College (Herries Davies, 1965). This collection was composed of Irish Carboniferous invertebrates, Jurassic ichthyosaurs, gastropods from the Pleistocene of Italy, a collection of volcanic and other rocks from continental Europe, zeolite minerals from the basalt of County Antrim, and some Irish rocks (M'Coy, 1841; Ball, 1848).

In 1857, the Museum collections were dispersed throughout the College campus: most of them, including the geological collections, were housed in a new Museum building erected for the purpose between 1854 and 1857 at a cost of 24,000 guineas. The hallway of this building contains columns and balustrades of nearly a dozen Irish marbles and elaborate floral and faunal decorative carvings (Caulfield, 2019; Wyse Jackson and Wyse Jackson, 2019).

Responsibility for the new Museum fell to James Apjohn (1796–1886), who held two chairs in Chemistry at the University and was Professor of Mineralogy from 1845 to 1881, and to Samuel Haughton (1821–1897), Professor of Geology from 1851 to 1881. It was a period of sparse acquisition by the Museum: specimens of particular interest include a collection of rocks from Napoleon's island prison of St. Helena; a collection of lavas from various eruptions of Vesuvius; and the main mass of the Dundrum meteorite, which fell in central Ireland in 1865.

In the 1950s, the Museum and its holdings were radically affected by a rationalisation programme. Space for teaching was required, and the Museum was removed to a room one-third of its former size (Wyse Jackson, 2019c). In addition, a number (probably considerable) of specimens were discarded; it is evident that these unfortunately included many specimens of immense historical interest.

Many specimens were incorporated into teaching collections, and labels were removed, making it difficult to work out the specimens' provenance and history. It was fortunate that the old handwritten catalogues were not also discarded.

The postgraduate research programme initiated in the 1960s has led to a rapid increase in the holdings of the Geological Museum. Most recent accessions are micropaleontological, but some macropaleontological, petrological, and occasional mineralogical material continues to find its way into the Museum collections. Occasionally specimens are purchased from mineral dealers: a number of meteorites were acquired in this way in 1990 and 1991.

At present, the Geological Museum contains approximately 72,000 specimens. These holdings include 7,000 mineralogical specimens, 15,000 petrological specimens, and about 50,000 palaeontological specimens. Approximately 60% of the specimens are adequately curated. Systematic cataloguing of the collections only began in the early 1960s. The Museum archives contain a number of early nineteenth-century handwritten catalogues of certain named collections: those of George Knox, Rev. George Graydon, and Robert Mallet (1810–1881)—the seismologist whose collection of Italian volcanic material is largely lost (Wyse Jackson, 1998b, 2007d). Unfortunately, it is now very difficult to correlate entries in these catalogues with extant specimens.

George Knox: parliamentarian and mineral collector

The Honourable George Knox (1765–1827), one-time Member of Parliament for Dublin University, was a member of a wealthy family which held considerable estates in Dungannon, County Tyrone. He was the fifth son of Thomas Knox, a Member of the Irish Parliament for Dungannon, who was later Baron Wells and then Viscount Northland. Thomas' sons gained high offices: Major-General John Knox (1758–1800) was also Member of Parliament for Dungannon and later became Governor of Jamaica; two other sons, William Knox (1762–1831) and Edmund Knox (1773–1849), became bishops of Derry and Limerick, respectively (Burke, 1849), William succeeding Frederick Hervey (1730–1803), Earl of Bristol and Bishop of Derry, the noted amateur vulcanologist and builder of large houses.

George Knox trained as a barrister at Lincoln's Inn, London, and was later called to the Irish Bar (Keane *et al.*, 1982). He received two honorary degrees in law, D.C.L. from Oxford (Foster, 1888) and LL. D. from Dublin (Todd, 1869).

While in London he became a close friend of Theobald Wolfe Tone and was a godfather to his son. Tone later organised an insurrection, with French aid, against the British authorities in 1798, and committed suicide after its failure. Prior to 1798, Knox had encouraged Tone to leave Ireland with his family, as his political activities were considered dangerous (Elliott, 1989).

Knox married twice and outlived both of his wives. He had five sons and one daughter (Burke, 1849). He lived at 8 Merrion Square East, a fashionable part of Dublin—probably, like many parliamentarians, he was well off. He served in the Irish Parliament as Member of the Borough of Dungannon from 1790 to 1797, and as Member of Dublin University from 1797 until the Act of Union of 1800 (Foster, 1888). He is noteworthy as an early campaigner for Catholic emancipation, which would have enfranchised the Roman Catholic majority of Ireland's population. Limited measures were passed in 1793 (Johnson, 1957), but Knox's motion that Roman Catholics should be eligible to sit in Parliament was thrown out (Elliott, 1989). Full emancipation came in 1829 following the championing of the cause by Daniel O'Connell.

In the late 1790s, the Irish Parliament was thrown into controversy over the attempts to disband it and create a union of the Dublin and London parliaments. Knox was opposed to the measure and resigned a government position in 1799 on account of it (Ingram, 1887; Malcolmson, 1978). The Act of Union was passed in 1800 and the Dublin Parliament was abolished; nevertheless, Knox continued to serve as the University Member of the Westminster Parliament in London until 1807, when he lost his seat in the general election of that year. Knox was also returned as Member of Dungannon in 1801 and 1805 by his father, in case he failed to be elected by the University. During his period in London, he was restored to favour and was appointed a Lord of the Treasury and a Privy Councillor (Ingram, 1887). Knox retired from Parliament in 1807, after which he served for two years as a Commissioner of Customs, receiving an annual pension of £400, and later in 1809, he was granted a pension of £800 per annum for life (Aspinall, 1986). He died at Velletri, 40 km south-east of Rome, on 13 June 1827 following an accident while on a journey to Naples where no doubt he wished to visit Vesuvius. The horses became startled and Knox fell from his carriage and broke his lower leg and femur. His headstone in the Non-Catholic cemetery at Caio Cestio in Rome recalls, in Latin, that he was an Irishman by birth and describes his injuries and how he lingered in considerable pain for fifty-three days before dying (Stanley-Price, 2024). It is perhaps surprising that the headstone doesn't recount his considerable public and scientific achievements, but focuses on his injury.

Why Knox served the University of Dublin in Parliament is something of a mystery, as prior to 1797 his only connections with the College were that he was suggested as a possible Provost (Head of the College) in 1792 (Aspinall, 1986), and that he was conferred with the honorary degree of Doctorate of Law in 1795. Perhaps, it was because of his service in Parliament that his mineral collection found its way into the College Museum.

Towards the end of the 1700s, Knox became active in scientific circles in Dublin and London and was a close friend of the mineralogist Sir Charles Lewis Giesecke (1761–1833) (Wyse Jackson, 1996b). Knox was elected a member of the Royal Society of London in 1802, and was a member of the Royal Irish Academy, the Dublin Society (Vice-President from 1820 to 1827), and the Kirwanian Society (President). The Kirwanian Society was established in 1812 to further the ideas of Richard Kirwan and to carry out mineralogical studies (McLaughlin, 1954). However, membership never rose above 77 and the Society disappeared after 1818. Knox contributed two papers to the Society: the first, on fiorin grass (a variety of *Agrostis stolonifera* Creeping Bent-Grass), was read on 21 March 1813, but was never published, while the second, which outlined the mineralogy of County Dublin for 'scientific travellers', appeared in the *Philosophical Journal and Scientific Review* long after the Society's demise (Knox, 1826). In it, Knox listed 43 mineral species and gave reasonably precise locality details for prospective collectors.

Knox was interested in analytical methods in mineralogy and carried out many experiments in the laboratories of the Royal Dublin Society. His results appeared in three papers: on the Calp Limestone of the Dublin region (Knox, 1802); on 'pitch-stone' from Newry, County Down (Knox, 1822); and on bitumen in rocks (Knox, 1823). In the first paper, published by the Royal Irish Academy, Knox described the distribution of the Lower Carboniferous argillaceous Calp Limestone and gave an analysis of it. Calp limestone was widely used as a building stone in Dublin until the late eighteenth century (Wyse Jackson and Caulfield, 2023). The meaning of the word 'Calp', first formally used by Richard Kirwan (1794), is unknown; it may have been used earlier as a quarrying term (Marchant and Sevastopulo, 1980).

In the second paper (1822), Knox outlined to the Royal Society of London his experimental methods for deriving bitumen from three different pitchstones. He distilled crushed stone to derive bitumen or bituminous water and an artificial pumice. He expanded this research by analysing 28 rocks and minerals for bitumen (Knox, 1823), and found that iron-bearing clays contained up to 18% bituminous water, while obsidian, tourmaline, augite, and hornblende were nearly devoid of bitumen. Concluding this work, Knox argued that previous mineral analyses were unreliable unless the rock or mineral had been distilled and the distillate analysed. And he postulated that the so-called 'Floetz' rocks (flat, stratified, fossiliferous rocks and associated basalts) of Werner were the chief source of the ejected volcanic rocks.

The Knox mineral collection

The Knox mineral collection was purchased by the University of Dublin for £500, a considerable sum at the time (Apjohn, 1850). The date of the purchase is unclear: Apjohn (1850) gives 1823, while Giesecke (1832b) states that the collection was purchased for the Museum 'after his [Knox's] lamented death'. Unfortunately, the minutes of the College Board for the years 1823–1830 do not clear this matter up, as they contain no mention of the collection. However, contained in the present holdings of the Museum are beryl specimens collected in the Mourne Mountains of County Down, probably in 1824 or 1825 (Giesecke, 1825b). It seems probable that the Knox collection was acquired after George Knox's death. Indeed, one would wonder why a zealous collector would sell his collection except for the money, and Knox was a man of some means who probably had no need of additional funds. It is unclear how many specimens were purchased by the College. A list at the beginning of the first catalogue states that the Knox collection numbered 2,398 specimens. However, considering that a catalogue of the Museum's minerals produced by Stokes (1818) five years prior to the Knox collection purchase lists 1,204 specimens, and the museum catalogue produced by Apjohn in 1850 lists only 1,991 specimens, the number of mineral specimens purchased from the Knox collection may not have been greater than 600. In addition, Giesecke and Sir Richard Griffith donated minerals to the Museum in the years between the publication of the two catalogues. Accompanying the Knox collection were three large catalogues in Knox's handwriting containing precise descriptions of specimens, locality details, and information about the physical properties of the mineral groups in general. The catalogues were compiled at various times; the first was made after 1814, the second (watermarked 1820) is a copy of the first; and the third (watermarked 1823) contains most of the information in the earlier two (Figure 21.1).

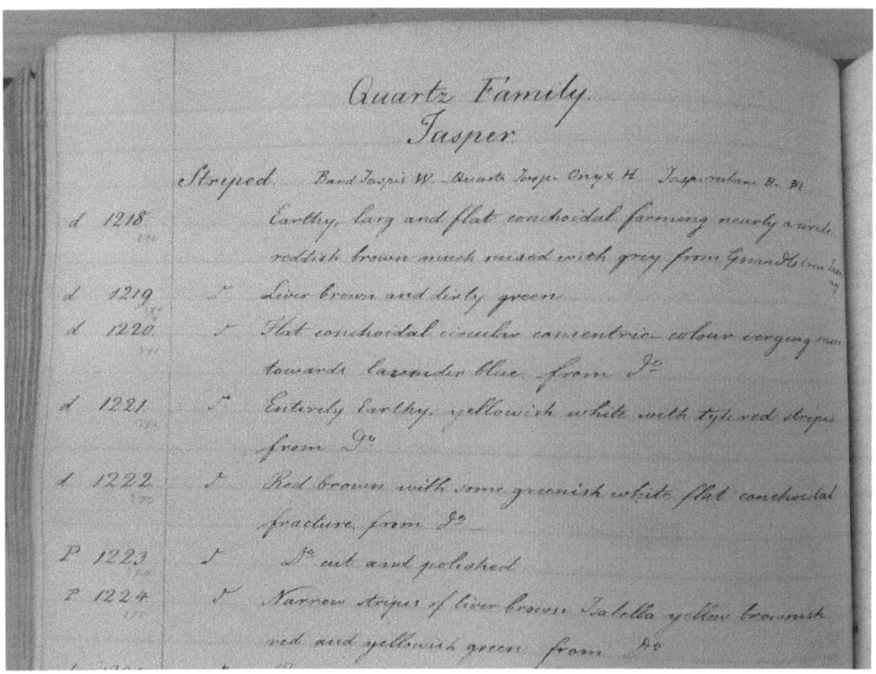
Quartz Family.
Jasper.
Striped.
d 1218. Earthy, large and flat conchoidal forming nearly a circle reddish brown much mixed with grey
d 1219. Do Liver brown and dirty green
d 1220. Do Flat conchoidal circular concentric colour varying towards lavender blue from Do
d 1221 Do Entirely Earthy yellowish white with red stripes from Do
d 1222 Do Red brown with some greenish white flat conchoidal fracture from Do
P 1223 Do Do cut and polished
P 1224 Do Narrow stripes of liver brown Isabella yellow brownish red and yellowish green from Do

Figure 21.1 Page of Knox Catalogue Number 3 with entry for 'Riband Jasper'. Knox number P1224 (Geological Museum, Trinity College Dublin).

Figure 21.2 'Riband Jasper', displaying alternating leek-green and brownish-red layers, 17.5 cm, from Saxony, Germany. George Knox collection. Modern catalogue number TCD.M2906.

Examination of Knox's catalogues reveals that many specimens were collected in the Americas (particularly the British colonies of the Caribbean), Brazil, Scandinavia, Siberia, Greenland, and Saxony (Figure 21.2). No doubt Knox's older brother John, the Governor of Jamaica, collected, or otherwise assisted in the acquisition of many of the Caribbean specimens, and a note in one of the catalogues indicated that some specimens from Greenland were given to Knox by Giesecke. A reasonable number were from Britain and Ireland, the Irish localities including the Giant's Causeway (Giesecke, 1825a), County Down (Giesecke, 1825b), and Dublin (Knox, 1826).

As a member of the Royal Irish Academy, Knox would have known Richard Kirwan, the prominent Irish mineralogist and chemist; indeed, the two men received their honorary degrees in Law within a year of each other. It is possible that Kirwan encouraged Knox and may have advised on the arrangement of the collection, as the catalogues are laid out according to the Wernerian principles which Kirwan promoted.

Specimens are placed into five classes (Table 21.1). The arrangement corresponds closely to the scheme laid out in Abraham Gottlob Werner's *Letztes Mineral-System* (1817), with a few exceptions: to Class 1 is added the genus Yttria, while Werner's genera called Barite and Strontian are reduced to familial level; in Class 4, the number of genera is greater (28, as against Werner's 22). Knox may have been the author of these modifications, but this is not stated in the catalogues. The constituents of Knox's fifth Class—*Colors*—are unlisted but may be related to categories in Werner's *Nomenclature of Colours* (Werner, 1814).

In 1850, Knox's mineral collection was incorporated with the Museum's pre-existing mineral collections by James Apjohn, who compiled a catalogue of the complete mineral holdings of the College in 1850. The specimens listed in this catalogue are today distinguished by small printed red numbers. Knox's specimens have distinctive handwritten labels bearing a number code corresponding to entries in his catalogue. Sadly, many of these labels were removed in the past. Nevertheless, it should be possible to recognise much of Knox's material.

Conclusion

The Knox collection, hitherto largely unknown outside of Trinity College, is important in terms of size and documentation, and in these senses, it compares favourably with other mineralogical collections of the 1700s and illustrates minor modifications of Werner's classification of minerals.

It is widely known that a small number of geologists in Ireland were at the forefront of the Neptunist–Vulcanist debate: men such as Kirwan and William Richardson (1740–1820) for the former, and Hamilton and Hervey for the latter, all took active roles. Although it is unclear what Knox's views were, it is probable that he sided with the Neptunists, given his involvement with the Kirwanian Society. This is the first piece of evidence of potential parliamentary input to the Neptunist-Vulcanist debate!

Table 21.1 Arrangement of the Knox Mineralogical Collection

Class 1. Earthy Substances [1,498 specimens], containing eight genera:

i. Diamond Genus	iv. Clay Genus	vii. Yttria Genus
ii. Zircon Genus	v. Talc Genus	viii. Hallite Genus
iii. Silica Genus	vi. Lime Genus	

Class 2. Salts [26 specimens], containing five genera:

i. Carbonatic Genus	iii. Nitric Genus	v. Sulphuric Genus
ii. Boratic Genus	iv. Muniatic Genus	

Class 3. Inflammables [63 specimens]

These are not listed in the catalogues.

Class 4. Metallic Substances [687 specimens], containing 28 genera:

i. Platina	xi. Lead	xx. Nickel
ii. Iridium	xii. Tin	xxi. Cobalt
iii. Osmium	xiii. Bismuth	xxii. Arsenic
iv. Rhodium	xiv. Zinc	xxiii. Molybdena
v. Palladium	xv. Cadmium	xxiv. Uranium
vi. Gold	xvi. Antimony	xxv. Scheelium
vii. Mercury	xvii. Tellurium	xxvi. Titanium
viii. Silver	xviii. Selenium	xxvii. Cromium
ix. Copper	xix. Manganese	xxviii. Cerium
x. Iron		

Class 5. Colours [124 specimens]

These are not listed in the catalogues.

22

SIR CHARLES LEWIS GIESECKE (1761–1833) AND GREENLAND: HIS MINERAL COLLECTION IN TRINITY COLLEGE, DUBLIN

Irish Journal of Earth Sciences 15(1996), 161–168.

Introduction

Charles Lewis Giesecke (1761–1833) was an important mineralogist who assembled large and unique collections of minerals from Greenland in the early 1800s. The history of the dispersal of these collections is well known, and the material can now be found in the Natural History Museum, London, the National Museum in Vienna, and in the Geological Museum of Copenhagen University. More material is possibly in the National Museum of Ireland, where it remains to be recognised. Recently, a long-forgotten suite of Giesecke's Greenland specimens, together with a manuscript catalogue of the same, has come to light in the Geological Museum, Trinity College, Dublin. This chapter describes the background to the acquisition, and the present extent of the collection.

Charles Lewis Giesecke

Sir Charles Lewis Giesecke (1761–1833) was born in Augsburg, Bavaria, Germany, on 6 April 1761, the eldest son of George Melchior Metzler, a Protestant tailor, and Sibylla Magadalena Gotz (Craig, 1993), and was named either Johann Georg or Karl Ludwig. Questions also surround the name and occupation of his father, which have been given as Johann Georg, a vintner (Petersen and Secher, 1993). Other sources suggest that Charles Lewis Giesecke may have been a twin (Scouler, 1834), and that he was the second-born son of the family (Monaghan, 1993). By the age of twenty, for reasons unknown, he had assumed the name Karl Ludwig Giesecke (anglicised after 1813 to Charles Lewis) by which he is now better known (Steenstrup, 1910). The problems of his name and that of his family are complex and need clarification. After studies in divinity and law at Gottingen, where he also took some courses in mineralogy under Johann Friedrich

DOI: 10.4324/9781003475552-27

Blumenbach (1752–1840), he embarked on a succession of careers, including those of musician, actor, diplomat, and mineralogist. In Vienna, where he had moved in the early 1780s, he published the opera Oberon, which was performed in 1790 (Craig, 1993), and appeared in the first performance of Mozart's *The Magic Flute* in 1791. He was to suggest later that he was the author of the libretto of the opera (Whittaker, 1991).

After a period of further study at Freiberg, under Abraham Gottlob Werner (1743–1817) in 1794, he became an Austrian diplomat, based in Constantinople and later in Italy. He left the diplomatic service on account of disability, and from 1801 spent several years engaged in mineralogical collecting trips. He eventually settled in Copenhagen, where he became a mineral dealer and from where, in 1805, he travelled to the Faeroe Islands on behalf of the Royal Greenland and Faeroese Trading Company (Monaghan 1993; Jørgensen, 1996). In 1806, he travelled to Greenland for the company in order to assess the mineral wealth and trading potential of the country (Monaghan, 1993; Whittaker, 2007). He was due to remain for three years but owing to the political unrest in Europe was unable to leave Greenland until 1813. There he assembled several mineral collections.

After a sojourn of seven years, Giesecke left Greenland on a whaling ship bound for Hull (Long, 1921a). On his arrival in England, he travelled to Scotland, where he met the mineral collector Thomas Allan (1777–1833), who had purchased some of his specimens that had been captured as a prize-of-war while in transit to Copenhagen. At Allan's suggestion Giesecke, who was then virtually penniless, applied for and was appointed to the chair of mineralogy at the Dublin Society (later the Royal Dublin Society), at a salary of £300 per annum (Mitchell, 1981). He was to hold this position for twenty years until his death in 1833. He was made a Knight Commander of the Royal Danish Order of Danneborg in 1816 when he was in Copenhagen sorting out his effects left there prior to his travels in Greenland. The Dublin Society recognised his service to it in 1817 by having a medal struck in both bronze and silver for him. By the celebrated Irish medallist William Stephen Mossop (1788–1827), it carries Giesecke's profile on its obverse side and a view of Disco Island, Greenland on the reverse side (Frazer, 1886, p. 463; Figure 22.1A, B).

For much of Giesecke life in Dublin he lived at 14 George's Place, now Hardwicke Place (Nigel Monaghan, pers. comm.). He died on 5 March 1833 and was buried in the adjoining churchyard of St George's Church, Hardwicke Place (Long, 1921a). A fine memorial to him was erected inside the church, it was executed by the sculptor Edward Smyth and carries a medallion of Giesecke in profile (Long, 1921b).

After his death, at which time he was said by his executor Robert Hutton to be living in abject poverty, his library was sold in the week ending 31 May 1834 in at least 1,311 lots, some of which were purchased by the English geologist George Bellas Greenough (Colain MacArthur, pers. comm.). His personal effects were retained by Hutton, including some autograph notebooks which were later described by Waterhouse in 1936. These were deposited in the National Museum

Figures 22.1A and B Charles Lewis Giesecke Medal struck in 1817. (A) Obverse with profile of Giesecke; (B) Reverse depicting Disco Island, Greenland with a polar bear, columnar basalt, icebergs, and a spouting whale (Author's collection).

of Ireland and in 1950 were donated to the National Library. A number of his manuscripts are now to be found in the National Library of Ireland (Anon., 1977), and may have been deposited with the Royal Dublin Society by Hutton's descendants (Waterhouse, 1936).

Giesecke's collections of minerals from Greenland

For seven years, between 1806 and 1813, Giesecke lived in Greenland, where he assembled large and valuable collections of minerals, as well as making important observations on the lives and traditions of the indigenous people, whose artefacts he also collected. He found a number of new mineral species, including Arfvedsonite (described by Brooke in 1823), Sapphirine (which he himself described in 1819), and Sodalite (described by Thomson in 1811), and he located the source of the rare mineral Cryolite at Ivikatt in Arsuk Fjord (Petersen, 1993).

During the Napoleonic wars in Europe, travel by sea was difficult, and ships were liable to be captured by the enemy. Giesecke's valuable collection of minerals, which was bound for Copenhagen, was captured by a French ship which in turn was taken by the English. The collection was acquired by the Scottish collector Thomas Allan for the paltry sum of £40. This material, which included valuable Cryolite specimens estimated to be worth at least £5,000 (Farrar and Farrar, 1968), is now to be found in the Natural History Museum, London. This institution purchased the collection from Robert Phillips Greg (1826–1906), whose father, Robert Hyde Greg (1795–1875), had purchased it from Allan's executors for £1,300 in 1833 (Cleevely, 1983).

When word reached Giesecke that the collection had been stolen, he assembled a second collection. Some of this material was sent to Copenhagen, where it was destroyed by enemy action in 1807 (Jørgensen, 1983). The majority of his second Greenland collection is probably in the National Museum of Ireland, and smaller collections are in Vienna and Copenhagen (donated in 1814) and in Trinity College Dublin.

On his appointment as professor of mineralogy to the Dublin Society in December 1813 (White, 1955) Giesecke donated 415 mineral specimens from Greenland to its museum (Scouler, 1834; Monaghan, 1993). He had to learn English and was soon delivering public lectures in the language. He was responsible for the already large mineralogical collections housed in the Society's museum on Hawkins Street (O'Riordan, 1983), which included those of Nathanael Leske, acquired in 1792. Soon he set out to enlarge the holdings of Irish minerals and did this by undertaking several collecting trips around his adopted country (Giesecke, 1826, 1828, 1829; MacArthur, 1987). In 1815, he arranged for the transfer of the collections to Leinster House, the new home of the Royal Dublin Society. The museum in this building consisted of six rooms; the Leskean collection was accommodated in one room, a collection of Irish material in another (called the 'Museum Hibernicum') and geological material in another. Giesecke's large collection of Greenland anthropological artefacts, such as a native hut and furniture, was on display in another room, together with his minerals. The museum was open to students at all times and to the general public for three hours from 12 noon on Mondays, Wednesdays and Fridays (Wright, 1821).

Giesecke also purchased minerals from and exchanged minerals with various European contacts, including Johann Wolfgang von Goethe (1749–1832) (Waterhouse, 1933) and the Imperial Museum in Vienna, which he visited between 1817 and 1819 (Hennig, 1951; Monaghan, 1993). These Viennese exchanges are listed in a manuscript now in the National Library of Ireland (MS 263) (Hennig, 1951). A comprehensive catalogue of the Royal Dublin Society mineralogical collection was published in 1832, the year before his death (Giesecke, 1832b). Subsequently, the collections were inherited by the National Museum of Ireland, where Giesecke's Greenland material has not at present been identified.

The Dublin University Museum and its mineralogical collections: Giesecke's additions

The Dublin University Museum was founded in 1777 to house among other collections, the Polynesian artefacts collected during the voyages of Captain James Cook (Wyse Jackson, 1992b). The driving force behind its establishment is thought to have been the Rev. William Hamilton (1755–1797), one-time fellow of the university, who is perhaps better known for his geological observations which advanced the idea of the igneous origin of the Giant's Causeway [Chapter 2 herein].

By 1807 the museum housed a collection of 1,089 minerals, some petrological specimens, including material from Vesuvius gathered by the cleric George Graydon (Vaccari and Wyse Jackson, 1995 [Chapter 3 herein]), and some fossils.

Whitley Stokes (1763–1845), lecturer in natural history, whose political activity brought him at one time into conflict with the university authorities, was responsible for the collections from the late 1790s until 1844. The mineralogical collections were enlarged through purchases and donations. Specimens were bought from dealers in Dublin, Derbyshire and Paris, and private collections were also acquired. Donations included minerals from Gascony given in 1798 by the professor of German (Wyse Jackson, 1994c) and specimens from Greenland donated by Giesecke in 1824.

A catalogue of the complete mineralogical collection was published in 1807 by Stokes, and in it, we find no specimens from Greenland listed. In the revised catalogue, published eleven years later, the collection included six Greenland specimens: Hyacinth (4), Iolite with Fahlunite (30), Sodalite from Kangertuarsuk (361), Plumose Anthophyllite from Ivertok (467), Cryolite from Ivikatt (749), and Allanite from Tindingen (1204) (Stokes, 1818—catalogue numbers in parenthesis). It is possible that these were given to the museum, or exchanged for specimens for the Royal Dublin Society museum, by Giesecke sometime after 1814

In 1823 a large collection of approximately 600 specimens was purchased for the considerable sum of £500 (£15,000 today) from the Hon. George Knox (1765–1827), one-time member of parliament for Dublin University [Chapter 21 herein]. In this one transaction, the university mineralogical collections were increased by a third to nearly 2,000 specimens. The Knox specimens include many from the Americas, particularly the British colonies of the Caribbean, and from Europe. Of particular interest are Beryl crystals from Ireland, particularly from the Mourne Mountains, said by Giesecke (1832) to be of exceptional quality. Equally of note, particularly in the context of the subject of this paper, are the specimens of Gieseckite from Greenland and Hornblende and Epidote from Norway given to Knox by Giesecke.

In 1823, Giesecke wrote to the board of Trinity College, Dublin, stating that he intended to present a suite of minerals, collected by him in Greenland, to the Dublin University Museum (TCD MS 846). He also stated that a catalogue of the collection would be forwarded at the same time. It is a measure of the value of the collection that the Board instructed the Registrar to write to Giesecke conveying their thanks, and that they granted him permission to use the library of the college. Such permission must have been very welcome to Giesecke, as the library contained the finest collections of books and manuscripts in Dublin.

Giesecke's catalogue is an eleven-page, unbound, coverless, handwritten document, on foolscap pages (330 mm × 203 mm in size) which bear the watermark of either a crowned lion clasping a trident within a crowned oval frame or the name and date 'J. Green 1824'.

Giesecke lists, in his distinctive and legible handwriting, in black ink (now faded to brown), 149 specimens, giving each a specimen number, the name of the material and a short description, and in most cases stating the location where the material was

Figure 22.2 Mineral specimens collected by Giesecke from Greenland: (A. Left) Cryolite from Arksutt; (B. Right) Hornblende from Kangerdluarsuk, with his distinctive hand-written specimen number label and small printed label (Geological Museum, Trinity College Dublin).

collected. This numbering scheme is very distinctive, with the first number 'to one decimal place' followed by a dash, after which is a one- to three-digit number (e.g. 6.1–32, or 1204.2–90). It is possible that this is a numbering scheme popularised by Werner. In most cases, the specimens are from Greenland, either from the continent or from some offshore islands. The material from Greenland was varied and included specimens of the rarer minerals, such as Allanite, Cryolite (listed as Kryolite; Figure 22.2A) and Sodalite, large numbers of zeolite minerals, and more common mineral species, including Garnet, Hornblende and Tourmaline. The catalogue also lists seventeen specimens, mainly of Tourmaline and Felspar, from Arendal, Bucen and Naes in Norway, and one specimen, Tourmaline with Rubellite, from North America (Table 22.1). Giesecke also included in his gift a lava specimen that he had procured from an iceberg from Iceland.

Using the catalogue, the collections in the Geological Museum were searched and 46 of Giesecke's specimens have now been recognised. In nearly all cases they bear two or three labels that are Giesecke's own and a further two labels that have been attached since 1826. Giesecke's consist of a small hand-written label on which the specimen number is written, and a smaller printed label (Figure 22.2B). On the latter, a number to one decimal place appears above a line underneath which is a one- to three-digit number. The number is written in black ink identical to that used in the catalogue, except for the digit after the decimal place, which is written in red ink. The black ink has faded to a brown colour, while the red ink in most cases has faded but is legible under ultraviolet light. The smaller printed label bears a printed number that matches the hand-written number beneath the line. In addition to the specimens listed in the catalogue are a small number that carry only the small printed label these may be the specimens given prior to 1824.

In 1850, small labels with red printed numbers were added, the numbers corresponding to entries in James Apjohn's catalogue (Apjohn, 1850). This catalogue lists only 41 specimens from Greenland. Perhaps the remainder of Giesecke's donation was held in a stratigraphic collection rather than in the systematic mineralogical holdings. In the 1970s, numbers prefixed with 'M' were given to specimens recatalogued at that time.

Table 22.1 List of Mineral Species, and Their Area of Origin, in the Giesecke Donation to Trinity College, Dublin. The Names are as spelt by Giesecke; the Modern Accepted Species Name Is Given in Italics.

Greenland	Copper pyrites
Actinolite	Diallage [probably *Hypersthene*]
Hornblende	Dichroite (Iolite) [*Cordierite*]
Hypersthene	Epidote
Indurated green earth	Floating mesotype
Sapphirine	Fluor [*Fluorite*]
Silver white mica [*Muscovite*]	Garnet
Sodalite	Granular limestone
Allanite	Indurated marl
Amethyst (baccillar white)	Inflammable shale
Analcime	Indurated talc
Apatite	Iron pyrites
Apophyllite	Kryolite [*Cryolite*]
Arctigite [*Scapolite*]	Labrador felspar [*Labradorite*]
Asbestus [*Asbestos*]	Massive iron pyrites
Aventurin felspar [*Sunstone—Oligoclase*]	Magnetic iron ore
Aventurin quartz [*Quartz*]	Mesotype [zeolite group name]
Baccillar arragonite	Stilbite
Blackish-brown lava	Syenite
Calcareous spar [*Calcite*]	Tourmaline
Chabasie [*Chabazite*]	Tremolite
Chalcedony	Wavellite
Coccolite [var. of *Diopside*]	Whitestone (compact felspar) [*Orthoclase*?]
Copper glance [*Chalcocite*]	Yellow rhomboidal spar [?]

(Continued)

Table 22.1 (Continued)

Norway	Precious garnet [*Pyrope* and *Almandine*]
Milk quarts [*Quartz*]	Precious apophyllite [*Apophyllite*]
Needlestone [*Mesotype*]	Precious hypersthene [*Hypersthene*]
Opalising adular felspar	Precious serpentine [*Serpentine*]
Augite	Red Kryolite [*Cryolite*]
Diopside [*Orthoclase*]	Rock crystal [*Quartz*]
Elceolite [?]	Sahlite
Pearl-white felspar	Hornblende
Epidote	Magnetic iron ore
Felspar	Scapolite
Garnet	Titanite
North America	
Tourmaline, with Rubellite	

REFERENCES

Ackroyd, W. 1901a. On the circulation of salt and its bearing on chemico-geological problems, more particularly that of the age of the Earth. *Chemical News* 83: 265–268.

Ackroyd, W. 1901b. On the circulation of salt in its relations to geology. *Geological Magazine*, New Series, Decade 4, 8: 445–449; 558–559.

Adelman, J. 2007. Eozoon: debunking the Dawn Animal. *Endeavour* 31: 94–98.

Allen, C. 1975. *Plain tales from the Raj. Images of British India in the twentieth century.* London: André Deutsch Ltd and the British Broadcasting Corporation.

Allen, D.E. 1976. *The Naturalist in Britain: a social history*. Princeton: Princeton University Press.

Allen, D.E. 1990. *Arcana ex multitudine*: prosopography as a research technique. *Archives of Natural History* 17: 349–359.

Andrews, J.H. 1969. Introduction. In Taylor, G. and Skinner, A. *Map of the roads of Ireland.* Shannon: Irish University Press, v–xv.

Andrews, M.K. 1893. Denudation at Cultra, Co. Down. *Proceedings of the Belfast Naturalists' Field Club* 3: 529–532.

Andrews, M.K. 1894. Dykes in Antrim and Down. *Irish Naturalist* 3: 93–96.

Andrews, M.K. 1901a. Notes on Moel Tryfaen. *Proceedings of the Belfast Naturalists' Field Club* 4: 205–210.

Andrews, M.K. 1901b. Erosion at Newcastle, Co. Down. *Irish Naturalist* 10: 114.

Anglesea, M. 1997. The art of nature illustration. In Wilson Foster, J. and Chesney, H. (eds.) *Nature in Ireland.* Dublin: Lilliput Press, 497–523.

Anglesea, M. and Preston, J. 1980. 'A philosophical landscape': Susanna Drury and the Giant's Causeway. *Art History* 3: 252–273.

Anon. 1794. A list of donations made to the Royal Irish Academy since the publication of the second volume of their Transactions, 1789. *Transactions of the Royal Irish Academy* 5: 1–4.

Anon. 1801. Minutes of the council. *Proceedings of the Dublin Society* 37: 122.

Anon. 1804. Minutes of the council. *Proceedings of the Dublin Society* 40: 34.

Anon. 1838. *Atlas to accompany 2d report of the Railway Commissioners Ireland.* Dublin.

Anon. 1862. Proceedings of meetings: Monday, February 27, 1860. *Proceedings of the Royal Irish Academy* 7: 276–280.

Anon. 1869. Charles Æmilius Oldham, B.A. *Geological Magazine*, 6: 240.

Anon. 1899. Review of 'An estimate of the Geological Age of the Earth', by J. Joly, University of Dublin. *American Journal of Science* 158, 390–392.

Anon. 1902. *The Belfast Literary Society 1801–1901.* Belfast: Belfast Literary Society.

Anon. 1972. William King. *Geological Survey of India News* 2(4): 6–9.

Anon. 1977. Appendix manuscripts in the National Library of Ireland not included in manuscript sources for the history of Irish civilisation or supplements. *Report of the Trustees of the National Library of Ireland,* 23–27. Dublin: Stationery Office

Apjohn, J. 1850. *A descriptive catalogue of the minerals in the Museum of Trinity College Dublin*. Dublin: M.H. Gill.

Archer, J.B. 1980. Patrick Ganly: geologist. *Irish Naturalists' Journal* 20: 142–148.

Archer, J.B. 1985. Patrick Ganly: geologist. In Mollan, R.C., Davis, W.J. and Finucane, B. (eds.) *Some people and places in Irish science and technology.* Dublin: Royal Irish Academy, 40–41.

Archer, J.B. 1987. Science loners: the *Journal of the Geological Society of Dublin* and its successors. In Hayley, B. and McKay, E. (eds.) *Three Hundred Years of Irish Periodicals.* Mullingar: Association of Irish Learned Journals and Lilliput Press, 49–68.

Archer, J.B. 1993. Geological artistry: the drawings and watercolours of George Victor du Noyer in the archives of the Geological Survey of Ireland. In Dalsimer, A.M. (ed.) *Visualizing Ireland: national identity and the pictorial tradition.* Boston and London: Faber and Faber, 133–144.

Archer, J.B. 1995. Field Folios: Du Noyer's outdoor geological artistry. In Croke, F. (ed.) *George Victor Du Noyer 1817–1869: hidden landscapes.* Dublin: National Gallery of Ireland, 49–63.

Arduino, G. 1782. Effetti di antichissimi estinti vulcani, ed altri fenomeni, e Prodotti Fossili os-servati [...] nella Villa di Chiampo ed in altri luoghi del Territorio di Vicenza [letter dated Chiampo, 12th March 1769]. *Nuovo Giornale d'Italia* 8: 161–167.

Arkell, W.J. and Tomkeieff, S.I. 1953. *English rock terms chiefly as used by Miners and Quarrymen.* Oxford University Press.

Aspinall, A. 1986. Knox, Hon. George (1765–1827) of Dungannon Park, Co. Tyrone. In Thorne, R.G. (ed.) *The history of parliament-the House of Commons 1790–1829.* Volume 4. London: Secker and Warburg, 346–348.

Austen, R.A.C. 1850. On the valley of the English channel. *Quarterly Journal of the Geological Society of London* 6: 69–97.

Badash, L. 1968. Rutherford, Boltwood, and the age of the earth; the origin of radioactive dating techniques. *Proceedings of the American Philosophical Society* 112(3), 157–169.

Baily, W.H. 1855. Description of some Cretaceous Fossils from South Africa, collected by Captain Garden. *Quarterly Journal of the Geological Society* 11: 454–465.

Baily, W.H. 1858–1859. On fossils from the Crimea, with descriptions of some new species, principally in the collection of the Royal Dublin Society. *Journal of the Royal Dublin Society* 2, 233–240.

Baily, W.H. 1858. Description of Fossil Invertebrata from the Crimea. *Quarterly Journal of the Geological Society* 14: 133–163.

Baily, W.H. 1859a. On the fructification of *Cyclopteris Hibernica*, Forbes, from the Upper Devonian or Lower Carboniferous strata at Kiltorkan Hill, county Kilkenny. *Report of the British Association for the Advancement of Science (Leeds, 1858).* London: John Murray, 75–76.

Baily, W.H. 1859b. On the occurrence of detached plates of the shell of a new species of Chiton in the Carboniferous Limestone at Lisbane, county of Limerick. *Natural History Review* 6: 330–334; *Journal of the Geological Society of Dublin* 8 (1860), 167–171.

Baily, W.H. 1860a. On *Sphenopteris Hookeri*, a new Fossil Fern from the Upper Old Red Sandstone formation at Kiltorkan Hill, in the county of Kilkenny; with some observations upon the Fish-remains and other associated fossils from the same locality. *Report of the British Association for the Advancement of Science (Aberdeen, 1859).* London: John Murray, 98–99.

Baily, W.H. 1860b. On a new species of *Solarium* from the Upper Greensand near Dorchester. *Dublin Zoological and Botanical Association Proceedings* 2: 66–67.

Baily, W.H. 1862. Remarks on some Coal-Measure Crustacea belonging to the genus *Belinurus*, König, with descriptions of two new species from Queen's County, Ireland. *Proceedings of the Dublin University Zoological and Botanical Association* 3: 88–96.

Baily, W.H. 1866. The recent discovery of fossil reptiles in the coal of the south of Ireland. *Geological Magazine* 3: 84–88.

Baily, W.H. 1867–1875. *Figures of characteristic British fossils: with descriptive remarks, volume 1, Palaeozoic, with 42 plates and 18 woodcuts.* London: Van Voorst. [Parts 1–4. Part 1, 1–30, 1867; Part 2, 31–61, 1869; Part 3, 62–92, 1871; Part 4, i–lxxx, 93–126, 1875].

Baily, W.H. 1869. Notes on the fossils from the Old Red Sandstone of Kiltorcan Hill, Co. Kilkenny. *Report of the British Association for the Advancement of Science (Norwich, 1868).* London: John Murray, 58–59.

Baily, W.H. 1870. Report on fossils obtained at Kiltorcan Quarry, Co. Kilkenny. *Report of the British Association for the Advancement of Science (Exeter, 1869).* London: John Murray, 73–75.

Baily, W.H. 1874. Sketch of the geology of Belfast and the neighbourhood. *Hardwicke's Science Gossip* 116: 169–170.

Baily, W.H. 1876. Description of a new species of Labyrinthodont Amphibian from the Coal at Jarrow Colliery, near Castlecomer, Kilkenny. *Report of the British Association for the Advancement of Science (Bristol, 1875).* London: John Murray, 62–63.

Baily, W.H. 1877. On fossils from the Upper Old Red Sandstone of Kiltorkan Hill, in the county of Kilkenny, Report No. 1. *Proceedings of the Royal Irish Academy* 2: 45–48.

Baily, W.H. 1878a. Palaeontology of County Dublin. In Macalister, A. and M'Nab, W.R. (eds.) *Guide to the county of Dublin: its geology, industries, Flora, and Fauna.* Dublin: Hodges, Foster & Figgis, 48–72. Reprinted in 1880 in the *Journal of the Royal Geological Society of Ireland* 15: 78–98 and the *Scientific Proceedings of the Royal Dublin Society* 1: 162–182.

Baily, W.H. 1878b. Sketch of the geology of Dublin and Wicklow. *Hardwicke's Science Gossip* 164: 179–183.

Baily, W.H. 1887. *Rambles on the Irish Coast, especially relating to its geology, natural history and antiquities.* Dublin: Sealy, Bryers and Walker.

Ball, R. 1846. *First report on the progress of the Dublin University Museum, January 1846.* Dublin: M.H. Gill.

Ball, R. 1847. *Second report on the progress of the Dublin University Museum, June 1847.* Dublin: M.H. Gill.

Ball, R. 1848. *The Dublin University Museum, December, 1848.* Dublin: M.H. Gill.

Ball, R. 1853. Evidence. In *Dublin University Commission, Report of Her Majesty's commissioners appointed to inquire into the state, discipline, studies and revenues of the University of Dublin and Trinity College.* Dublin: Alexander Thom, 153–169.

Ball, V. 1880. *Jungle life in India: or the journeys and journals of an Indian geologist.* London: Thomas de la Rue.

Bancroft, H.H. 1893. *The book of the fair.* Chicago and San Francisco: The Bancroft Company.

Barber, L. 1980. *The heyday of natural history, 1820–1870.* New York: Doubleday.
Barrell, J. 1917. Rhythms and the measurements of geologic time. *Geological Society of America Bulletin* 28: 745–904.
Bassett, M.G. 1975. Bibliography and index of catalogues of type, figured and cited fossils in museums in Britain. *Palaeontology* 18: 753–773.
Bateman, J., McKenna, G. and Timberlake, S. (eds.) 1993. *Register of natural science collections in South East Britain.* Cambridge: Museum Documentation Association.
Batson, R.M. and Russell, J.F. (eds.) 1995. Gazetteer of planetary nomenclature 1994. *US Geological Survey Bulletin* 2129.
Becker, G.F. 1908. Age of a cooling Globe in which the initial temperature increases directly as the distance from the surface. *Science* 27(684): 227–233.
Becker, G.F. 1910a. Halley on the age of the ocean. *Science* 31: 459–461.
Becker, G.F. 1910b. Reflections on J. Joly's method of determining the ocean's age. *Science* 31: 509–512.
Becker, G.F. 1910c. The age of the Earth. *Smithsonian Miscellaneous Collections* 56(6): 1–28.
Beduant, F.S. 1822. *Voyage minéralogique et géologique, en Hongrie, pendant l'année 1818. Relation historique 1822.* Volume 4. Paris: Verdière.
Benton, M.J. and Taylor, M.A. 1983. Marine coast reptiles from the Upper Lias (Lower Toarcian, Lower Jurassic) of the Yorkshire. *Proceedings of the Yorkshire Geological Society* 44: 399–429.
Berg-Madsen, V. 1989. Origin and usage of the geological terms orsten, stinkstone and anthraconite. *Archives of Natural History* 16: 191–208.
Berger, J.-F. 1816. On the geological features of the north-eastern counties of Ireland. *Transactions of the Geological Society of London* 3: 121–195.
Berger, J.F. and Conybeare, W. 1816. On the geological features of the north-eastern counties of Ireland. *Transactions of the Geological Society of London* 3: 121–222.
Berry, H.F. 1915. *A history of the Royal Dublin Society.* London: Longmans, Green and Co.
Blackstock, A. 2013. *Science, politics and society in early nineteenth-century Ireland: the Reverend. William Richardson.* Manchester: University of Manchester Press.
Blot, J. 1969. *Les poissons fossiles du Monte Bolca classes jusqu'ici les families des Garangidae, Menidae, Ephippidae, Scatophagidae.* Verona: Studi e Ricerche sui giacimenti terziari di Bolca. Volume 1.
Bozza, V. 1788. *Dell'universale rivoluzione sofferta dal globo terracqueo.* Verona: Lettera al D. Orazio Rota.
Brice, W.R. 1982. Bishop Ussher, John Lightfoot and the age of creation. *Journal of Geological Education* 30: 18–24.
Bright, K. 2004. *The Royal Dublin Society, 1815–45.* Dublin: Four Courts Press.
Bright, R. 1818. *Travels from Vienna through Lower Hungary; with some remarks on the state of Vienna during the Congress in the Year 1814.* Edinburgh: Archibald Constable.
Brownrigg, W.B. 1865. Notes on part of the Leinster Coal Field, with a record of some fossils therein. *Journal of the Royal Geological Society of Ireland* 1: 145–147.
Brush, S.G. 1982. Finding the age of the earth by physics or by faith. *Journal of Geological Education* 30: 34–58.
Bryce, J. 1831. *Tables of simple minerals, rocks and shells; with local catalogues of species, for the use of the students of natural history in the Belfast Academy.* Belfast: Simms and M'Intyre.
Buckman, S.S. 1898. Observations of a cycle tour. *Proceedings of the Cotteswold Naturalists' Field Club* 12: 217–238.

Bulkeley, R. 1693. Part of a letter from Sir R.S. S.R.S. to Dr. Lister, concening the Giant's Causeway in the County of Atrim in Ireland. Old Bawn Apr. 24, 1693. *Philosophical Transactions of the Royal Society of London* 17: 708–710.

Bullard, E.C. and Gaskell, T.F. 1941. Submarine seismic investigations. *Proceedings of the Royal Society of London* 177A: 476–499.

Buntebarth, G. 2002. Temperature measurements below the Earth's surface: a history of records. *Earth Sciences History* 21: 190–198.

Burchfield, J.D. 1990. *Lord Kelvin and the age of the Earth.* 2nd edition. Chicago: University of Chicago Press.

Burek, C.V. 2007. The role of women in higher education Bedford College and Newnham College. In Burek, C.V. and Higgs, B. (eds.) *The role of women in the history of geology.* Geological Society, London, Special Publications 281: 9–38.

Burke, J.B. 1849. *A genealogical and heraldic dictionary of the peerage and baronetcy of the British Empire.* London: Henry Colburn.

Burns, D.T. 1977. Irish contributions to analytical Chemistry. *Proceedings of the Analytical Division of the Chemical Society* 14: 171–177.

Burns, D.T. 1985. Richard Kirwan: chemist, mineralogist, geologist and meteorologist, 1733–1812. In Mollan, R.C., Davis, W.J. and Finucane, B. (eds.) *Some people and places in Irish science and technology.* Dublin: Royal Irish Academy, 14–15.

Burns, D.T. 2003. *Richard Kirwan, 1735–1812: "The philosopher of Dublin".* Dublin: Royal Irish Academy.

Burns, V. and Rickards, R.B. 1993. Silurian graptolite faunas of the Balbriggan Inlier, counties Dublin and Meath, and their evolutionary, stratigraphical and structural significance. *Proceedings of the Yorkshire Geological Society* 49: 283–291.

Burstyn, H.L. 1960. The Salem philosophical library: its history and importance for American science. *Essex Institute Historical Collections* 96: 169–206.

Butler, G.W. 1889. On the subdivision of the body-cavity in lizards, crocodiles, and birds. *Proceedings of the Zoological Society of London* 57: 452–474.

Butler, G.W. 1892. Abstract of Mr. A. Ricco's account of the submarine eruption north-west of Pantellaria, October 1891. *Nature* 45(1173): 584–585.

Butler, G.W. and Perry, G.H. 1892. On the matter thrown up during the eruption north-west of Pantelleria, October 1891. *Nature* 45(1159): 251–253.

Capuano, G. 1994. *Viaggiatori britannici a Napoli tra '500 e '600.* Salemo: Pietro Laveglia Editore.

Capewell, J.G. 1965. The Old Red Sandstone of Slieve Mish, Co. Kerry. *Proceedings of the Royal Irish Academy* 64B: 165–174.

Carte, A. and Baily, W.H. 1863. Description of a new species of *Plesiosaurus*, from the Lias, near Whitby, Yorkshire. *Journal of the Royal Dublin Society* 4: 160–170.

Caulfield, L.M. 2019. The Irish marble industry and the Museum Building. In Casey, C. and Wyse Jackson, P.N (eds.) *The Museum Building of Trinity College Dublin: setting the standard in design, materials and craftsmanship.* Dublin: Four Courts Press, 13–51.

Chamberlin, T.C. 1899. Lord Kelvin's address on the age of the Earth as an abode fitted for life. *Science* 10(236): 11–18.

Chamberlin, T.C. 1922. The age of the earth from the geological viewpoint. *Proceedings of the American Philosophical Society* 61(4): 247–271.

Chaudhuri, N.P. 2001. *Story of GSI 1851–2002.* Calcutta: Geological Survey of India.

Chester, A.H. 1896. *A dictionary of the names of minerals including their history and etymology.* New York and London: John Wiley and Sons and Chapman and Hall.

Childe-Pemberton, W.S. 1925. *The Earl Bishop: the life of William Hervey, Bishop of Derry, Earl of Bristol.* 2 volumes. London: Hurst and Blackett.

Ciancio, L. 1995a. The correspondence of a virtuoso of the late enlightenment: John Strange and the relationship between British and Italian naturalists. *Archives of Natural History* 22: 119–129.

Ciancio, L. 1995b. *Autopsie della Terra. Illuminismo e geologia in Alberto Fortis (1741–1803).* Firenze: Leo S. Olschki.

Clarke, A.M. 1993. *Hey's mineral index: mineral species, varieties and synonyms*. London: Chapman and Hall.

Clarke, F.W. 1910. A preliminary study of chemical denudation. *Smithsonian Miscellaneous Collections* 56(5): 1–19.

Cleevely, R.J. 1983. *World palaeontological collections*. London: British Museum (Natural History) and Mansell Publishing Ltd.

Coffey, P. 1993. George Victor Du Noyer 1817–1869. *Sheetlines* 35: 14–26.

Colby, T.F. 1837. *Memoir of the city and north western liberties of Londonderry. Parish of Templemore. Ordnance Survey of Londonderry.* Volume the first. Dublin: Hodges and Smith for Her Majesty's Government.

Cole, G.A.J. 1887. The rhyolites of Wuenheim, Vosges. *Geological Magazine,* 24: 181–182.

Cole, G.A.J. 1892a. County Dublin, past and present. *Irish Naturalist* 1: 9–14, 31–36, 53–57, 73–76, 90–95. [Reprinted 1892 with different pagination as *County Dublin, Past and Present*. Dublin: Webb & Sons].

Cole, G.A.J. 1892b. The variolite of Annalong, Co. Down. *Scientific Proceedings of the Royal Dublin Society* 7: 112–116.

Cole, G.A.J. 1894a. Geologists at the Luncheon Table. *Irish Naturalist* 4: 41–44.

Cole, G.A.J. 1894b. On variolite and other tachylytes at Dunmore Head, Co. Down. *Geological Magazine* 31: 220–222.

Cole, G.A.J. 1894c. *The Gypsy Road: a journey from Krakow to Coblentz.* London and Mew York: Macmillan & Co..

Cole, G.A.J. 1895a. *Open-air-studies: an introduction to geology out-of-doors*. London: Griffin and Co.

Cole, G.A.J. 1895b. *Scenery and geology in County Antrim.* Belfast: R. Carswell and Son.

Cole, G.A.J. 1896. The rhyolites of the County of Antrim. *Scientific Transactions of the Royal Dublin Society* 6: 7–114.

Cole, G.A.J. 1897. On the geology of Slieve Gallion, in the county of Londonderry. *Scientific Transactions of the Royal Dublin Society* 6: 213–248.

Cole, G.A.J. 1898a. The floor of a continent. *Knowledge* 21: 25–28.

Cole, G.A.J. 1898b. Structure of Ireland. *Knowledge* 21: 74–78.

Cole, G.A.J. 1898c. The Mourne mountains. *Knowledge* 21: 121–124.

Cole, G.A.J. 1898d. An old-world highland. *Knowledge* 21: 170–173.

Cole, G.A.J. 1898e. An Esker in the plain. *Knowledge* 21: 217–220.

Cole, G.A.J. 1898f. Volcanoes in the north. *Knowledge* 21: 266–268.

Cole, G.A.J. 1911a. The Lost City [The King (1158); The City (1450); The Watchman (1500); The City (1911); The Causse du Larzac (1911)]. *Irish Review* 1: 322–324.

Cole, G.A.J. 1911b. Glacial features in Spitsbergen in relation to Irish geology. *Proceedings of the Royal Irish Academy* 29B: 191–208.

Cole, G.A.J. 1912a. *The growth of Europe*. London: Williams and Norgate.

Cole, G.A.J. 1912b. *Ireland: the land and the landscape*. Dublin: Educational Company of Ireland.
Cole, G.A.J. 1912c. The problem of the Liffey Valley. *Proceedings of the Royal Irish Academy* 30B: 8–19.
Cole, G.A.J. 1916. On the mode of occurrence and origin of the orbicular granite of Mullaghderg, Co. Donegal. *Scientific Proceedings of the Royal Dublin Society* 15: 141–158.
Cole, G.A.J. 1918a. On lecturing with the Lantern. *Department of Agriculture and Technical Instruction for Ireland Journal* 19: 14–24.
Cole, G.A.J. 1918b. *On lecturing with the lantern*. Dublin: T.H. Mason.
Cole, G.A.J. 1919. *Ireland the outpost*. Oxford: Oxford University Press.
Cole, G.A.J. 1921a. Why we teach geography. *Journal of Education and the School World* 53: 25–27.
Cole, G.A.J. 1921b. Why do we not teach geography? *The Geographical Teacher* 10: 276–279.
Cole, G.A.J. 1924. *Entry in Who's Who*. London: A. & C. Black, 581.
Cole, G.A.J. and Butler, G.W. 1892. On the lithophyses in the obsidian of the Rocce Rosse, Lipari. *Quarterly Journal of the Geological Society of London* 48: 438–446.
Cole, G.A.J. and Cole, B. 1902. *As We Ride*. Dublin: Royal City of Dublin Hospital.
Cole, G.A.J. and Crook, T. 1910. *On rock-specimens dredged from the floor of the Atlantic of the Coast of Ireland, and their bearing on submarine geology*. Dublin: Geological Survey of Ireland.
Cole, G.A.J. and Gregory, J.W. 1890. The variolitic rocks of Mt. Genevre. *Geological Magazine* 24: 140–141.
Collie, M. and Deimer, J. (eds.) 2004. *Murchison's wanderings in Russia*. Nottingham: British Geological Survey.
Collins, C., Cooper, J.A., Roden, R., Simmons, M., Timberlake, S. and Thompson, S. 1996. The BCG/GCG orphan collections working party report. *The Geological Curator* 6: 271–273.
Collins, G.B. 1976. *Wildcats and shamrocks*. North Newton, Kansas: Mennonite Press.
Comparato, V.I. 1979. Viaggiatori inglesi in Italia tra Sei e Settecento: la formazione di un mo-dello interpretativo. *Quaderni Storici* 42: 850–866.
Conybeare, W. and Buckland, W. 1816. Descriptive notes referring to the outline of sections presented by a part of the coasts of Antrim and Derry. *Transactions of the Geological Society of London* 3: 196–216.
Conybeare, W. and Berger, J.-F. 1816. On the geological features of the North-eastern counties of Ireland. *Transactions of the Geological Society of London* 3: 121–222.
Cook, A. 1998. *Edmond Halley: charting the heavens and the seas*. Oxford: Clarendon.
Cooke, J. (undated). *John Boyd Dunlop*. Dreoilín.
Cooke, J. 1902. *Murray's handbook for Ireland*. 6th edition. London: Edward Stanford.
Coonan, S. 1991. John Joly's Irish tricolour. *Technology Ireland* 23(6): 18–21.
Cotton, H. 1848. *Fasti Ecclesiae Hibernicae: the province of Leinster*. Dublin: Hodges and Smith, vol. 2, 452 p.
Cox, L.R. 1942. New light on William Smith and his work. *Proceedings of the Yorkshire Geological Society* 25: 1–99.
Cox, L.R. 1948. *William Smith and the birth of stratigraphy*. London: International Geological Congress, 1–8.

Cox, R. (ed.) 1982. *Robert Mallet, F.R.S., 1810–1881*. Dublin: Institution of Engineers of Ireland and the Royal Irish Academy.

Cox, R. 1985. Robert Mallet: seismologist, materials scientist and engineer, 1810–1881. In Mollan, R.C., Davis, W.J. and Finucane, B. (eds.) *Some people and places in Irish science and technology*. Dublin: Royal Irish Academy, 42–43.

Craig, M.J. 1993. Charles Lewis Giesecke (1761–1833). In Nicholls, C.S. (ed.) *Dictionary of National Biography missing persons*. Oxford: Oxford University Press, 250–251.

Crane, M.D. 1983. Samuel Stutchbury (1798–1859), Naturalist and Geologist. *Notes and Records of the Royal Society of London* 37: 189–200

Crayshaw, L.R. 1908. On rock remains in the bed of the English channel. An account of the dredgings carried out by the SS 'Oithona' in 1906. *Journal of the Marine Biological Association of the United Kingdom* 8: 99–117.

Creese, M.R.S. and Creese, T.M. 1994. British women who contributed to research in the geological sciences in the nineteenth century. *British Journal of the History of Science* 27: 23–54.

Creese, M.R.S. and Creese, T.M. 2004. *Ladies in the laboratory II: West European women in science, 1800–1900*. Lanham, MD: Scarecrow Press, 49–51.

Croke, F. (ed.) 1995. *George Victor Du Noyer (1817–1869). Hidden landscapes*. Dublin: National Gallery of Ireland.

Cullen, F. 2004. Smith, Stephen Catterson. *Oxford Dictionary of National Biography*. Oxford: Oxford University Press.

Curry, D., Hamilton, D. and Smith, A.J. 1970. Geological and shallow subsurface geophysical investigations in the Western Approaches to the English channel. *Institute of Geological Sciences Report* 70(3): 1–12.

Dalrymple, G.B. 2001. The age of the Earth in the twentieth century: a problem (mostly) solved. In Lewis, C.L.E. and Knell, S.J. (eds.) *The age of the earth: from 4004 BC to AD 2002*. Geological Society, London, Special Publications 190: 205–221.

Dalton, C. 1890. *The Waterloo roll call*. London: Clowes and Sons.

Dalton, C. 1904. *The Waterloo roll call with biographical notes and anecdotes* (2nd edition). London: Eyre & Spottiswoode.

Dangeard, L. 1929. Observations de geologie sous-marine et d'oceanographie relatives a la Manche. *Annales de l'Institut Ocianographique*, New Series, 6: 1–295.

Davis, J.W. 1888. Catalogue of the collection of fossil fishes in the Science and Art Museum. Science and Art Museum, Dublin. 20 p.

Davis, P. and Brewer, C. 1986. *Register of natural science collections in North-East England*. Newcastle: North of England Museum Service.

Davis, W.M. 1912. A geographical pilgrimage from Ireland to Italy. *Annals of the Association of American Geographers* 2: 73–100.

Davis, W.J. 1985. The Royal Cork Institution. In Mollan, R.C., Davis, W.J. and Finucane, B. (eds.) *Some people and places in Irish science and technology*. Dublin: Royal Irish Academy, 104–105.

Davis, W.J. 2002. Thomas Andrews. In Mollan, R.C., Davis, W.J. and Finucane, B. (eds.) *Irish innovators in science and technology*. Dublin: Royal Irish Academy, 114–115.

Dean, D.R. 1991. Robert Mallet and the founding of seismology. *Annals of Science* 48: 39–67.

Dean, D.R. 1992. *James Hutton and the history of geology*. Ithaca: Cornell University Press

Deane, A. 1924. *The Belfast natural history and philosophical society centenary volume 1821–1921*. Belfast: Belfast Natural History and Philosophical Society.

DeArce, M., Monaghan, N.T. and Wyse Jackson, P.N. 2011. The uneasy correspondence between T.H. Huxley and E.P. Wright on fossil vertebrates found in Jarrow, Co. Kilkenny (1865–1867). *Notes and Records of the Royal Society* 65: 253–271.

DeArce, M. and Wyse Jackson, P.N. 2014. The alternate geologist: William King (1809–1886) and his scientific controversies. *Journal of the Galway Archaeological and Historical Society* 66: 100–124.

Deibert, J.E. and Breithaupt, B.H. 2016. *Tracks, trails, and thieves: the adventures, discoveries, and historical significance of Ferdinand V. Hayden's 1868 Geological Survey of Wyoming and adjacent territories*. Geological Society of America Special Paper 521.

Deisler, V.K. and Bassett, M.G. 1997. Bibliography and index of catalogues of type, figured and cited fossils in museums in Great Britain and Ireland (Supplement 1975–1996). *Palaeontology* 40: 597–617.

den Tex, E. 1996. Clinchers of the basalt controversy: empirical and experimental evidence. *Earth Sciences History* 15: 37–48.

Desmarest, N. 1768. Basalte d'Auvergne. *Encyclopedie. Recueil de Planches, sur les Sciences, les Arts Liberaux, et les Arts Mechaniques avec leur Explication.* Volume 4. Paris: Briasson, David, Le Breton.

Desmarest, N. 1771. Sur l'origine et la nature du basalte a grandes colonnes polygonales, determines par l'histoire naturelle de cette pierre, observe en Auvergne. *Mémoires de l'Académie des Sciences* 87: 705–775.

Desmond, A. 1994. *Huxley: the devil's disciple.* London: Michael Joseph.

Desmond, A. 1997. *Huxley: evolution's high priest.* London: Michael Joseph.

Diemer, J.A. 1996. Old or New Red Sandstone? Evolution of a nineteenth century stratigraphic debate, northern Scotland. *Earth Sciences History* 15: 151–166.

Dixon, F.E. 1971. Richard Kirwan: the Dublin philosopher. *Dublin Historical Record* 24: 53–64.

Dixon, H.H. 1934. John Joly 1857–1933. *Obituary Notices of the Royal Society of London* 3: 259–286.

Dixon, H.H. 1941. *John Joly: presidential address to the Dublin University Experimental Science Association*. Dublin: Dublin University Press.

Dixon, H.H. and Joly, J. 1895. On the Ascent of Sap. *Philosophical Transactions of the Royal Society of London B* 186: 563–576.

Dixon, H.H. and Joly, J. 1897. Coccoliths in our coastal waters. *Nature* 56: 468.

Dixon, H.H. and Joly, J. 1898. On some minute organisms found in the surface waters of Dublin and Killiney Bays. *Scientific Proceedings of the Royal Dublin Society* 8: 741–752.

Donovan, D.T. 1967. Henry Marc Brunel: the first submarine geological survey and the invention of the gravity corer. *Marine Geology* 5: 5–14.

Donovan, M. 1850. Biographical account of the late Richard Kirwan, Esq., President of the Royal Irish Academy, &c. *Proceedings of the Royal Irish Academy* 4: Appendix 8: lxxxi–cxviii

Dott, R.H. Jr. 1996. Lyell in America—his lectures, field work, and mutual influences, 1841–1853. *Earth Sciences History* 15: 101–140.

Doughty, P.S. 1980. Collections and collectors of note 33: James M'Donnel, M.D. (1762–1854). *Newsletter of the Geological Curators' Group* 2: 465–466.

Doughty, P.S. 1981. The state and status of geological collections in the UK. *Geological Society Miscellaneous Paper* 13: 1–118.

Drummond, W.H. 1811. *The Giants' Causeway, a poem*. Belfast: Longman, Hurst, Rees, Orme and Browne.

Elliott, M. 1989. *Wolfe Tone: prophet of Irish independence*. New Haven: Yale University Press.

England, P., Molnar, P. and Richter, F. 2007. John Perry's neglected critique of Kelvin's age for the Earth: a missed opportunity in geodynamics. *GSA Today* 17(1): 4–9.

Erskine, F.J. 1884. *Tricycling for ladies: or, hints on the choice & management of tricycles: with suggestions on dress, riding & touring*. London: Iliffe and Son.

Erskine, F.J. 1896. *Bicycling for ladies*. London: Iliffe and Son.

Erskine, F.J. 1897. *Lady cycling*. London: Walter Scott.

Evans, C.D.R. 1990. *The geology of the western English Channel and its western approaches, United Kingdom Offshore Regional Report*. London: HMSO.

Evans, E.E. and Turner, B.S. 1977. *Ireland's Eye: the photographs of Robert John Welch*. Belfast: Blackstaff Press.

Eve, A.S. 1939. *Rutherford*. New York: Macmillan.

Eyles, J.M. 1967. William Smith: the sale of his geological collection to the British Museum. *Annals of Science* 23: 177–212.

Eyles, J.M. 1969a. William Smith (1769–1839): a bibliography of his published writings, maps and geological sections, printed and lithographed. *Journal of the Society for the Bibliography of Natural History* 5: 87–109.

Eyles, J.M. 1969b. William Smith: some aspects of his life and word. In Schneer, C.J. (ed.) *Towards a history of geology*. Cambridge, MA and London: MIT Press, 142–158.

Eyles, J.M. 1972. Fitton, William Henry. *Dictionary of Scientific Biography* 5: 13–15.

Eyles, V.A. 1961. Sir James Hall, Bt. (1761–1832). *Endeavour* 20: 210–216.

Eyles, V.A. and Eyles, J.M. 1938. On the different issues of the first geological map of England and Wales. *Annals of Science* 3: 190–212.

Fara, P. 2005. Scientific coats of arms. *Endeavour* 29: 101–103.

Farrar, W.V. and Farrar, K.R. 1968. Thomas Allan, mineralogist an autobiographical fragment. *Annals of Science* 24: 115–120.

Farrington, A. 1954. Louis Bouvier Smyth. *Proceedings of the Geologists' Association* 65: 90–91.

Fermor, L.L. 1976. First twenty-five years of the geological survey of India. *Geological Survey of India Miscellaneous Publication* 39: 1–163.

Fisher, O. 1895. On the age of the world, as depending on the condition of the interior. *Geological Magazine*, New Series, Decade 4, 2: 244–246.

Fisher, O. 1900. Review of 'An estimate of the Geological Age of the Earth', by John Joly. *Geological Magazine*, New Series, Decade 4, 7: 124–132.

Fitton, W.H. 1824. Inquiries respecting the geological relation of the beds between the Chalk and the Purbeck Limestone in the south-east of England. *Annals of Philosophy* 24(8): 365–458. [Reprinted 1833. London: Richard Taylor].

Fitton, W.H. 1836. Observations on some of the strata between the Chalk and the Oxford Oolite, in the south-east of England. *Transactions of the Geological Society of London*, Series 2, 4: 103–389.

Fitton, W.H. 1852. Reply on award of the Wollaston Medal. *Proceedings of the Geological Society of London for 1852*: xx–xxi.

Flower, R.H. and Berdan, J.M. 1977. Memorial to Helen M. Duncan 1910–1971. *Geological Society of America Memorials* 5: 3.

Foley, S. 1694. An account of the Giant's Causeway in the North of Ireland. *Philosophical Transactions of the Royal Society of London* 18: 170–175.

Foote, R.B. 1868. On the distribution of stone implements in southern India. *Quarterly Journal of the Geological Society* 24: 484–495.

Forbes, E. 1854. Award of the Wollaston Medal and donation fund. *Proceedings of the Geological Society of London* 10: xix–xx.

Fortis, A., 1786. Extrait d'une lettre de M. l'Abbe Fortis datee de Veronne 24 septembre 1785 a M. le Comte de Cassini de l'Academie des Sciences sur les differentes petrification. *Journal de Physique* 28(1): 161–168.

Fortis, A. 1793. *Tre lettere sopra i pesci fossili di Bolca*. Venezia: A. Zatta e figli.

Foster, J. 1888. *Alumni Oxoniensis 1715–1886*. Volume 2. Oxford.

Fraser, R. 1801. *General view of the agriculture and mineralogy, present state and circumstances of the County Wicklow, with observations on the means of their improvement*. Dublin: Dublin Society.

Frazer, W. 1886. The medallists of Ireland and their work. *Journal of the Royal Historical and Archaeological Association of Ireland*, Series 4, 7: 443–466.

Fuller, J.G.C.M. 1995. *'Strata Smith' and his stratigraphic cross sections*. Tulsa and London: AAPG and Geological Society of London.

Fuller, J.G.C.M. 2001. Before the hills in order stood: the beginning of the geology of time in England. In Lewis, C.L.E. and Knell, S.J. (eds.) *The age of the earth: from 4004 BC to AD 2002*. Geological Society, London, Special Publications 190: 15–23.

Gabbey, W.A. 1985. William Molyneux: astronomer and natural philosopher, 1656–1698. In Mollan, R.C., Davis, W.J. and Finucane, B. (eds.) *Some people and places in Irish science and technology*. Dublin: Royal Irish Academy, 10–11.

Ganly, P. 1856. Observations on the structure of strata. *Journal of the Geological Society of Dublin* 7: 164–167.

Gasparini, P. and Musella, S. 1991. *Un viaggio al Vesuvio. Il Vesuvio visto attraverso diari, lettere e resoconti di viaggiatori*. Napoli: Liguori.

Gazola, G.B. and Tommaselli, G. 1794. *Lettere recentemente pubblicate sui pesci fossil veronesi con annotazioni inedite agli estratti delle medesime*. Verona: Ramanzini.

Geikie, A. 1875. *Life of Sir Roderick I. Murchison*. 2 volumes. London: John Murray.

Geikie, A. 1900. *Address to the geological section. Report of the 69th meeting of the British Association for the Advancement of Science (Dover, 1899)*. London: John Murray.

Geikie, A. 1924. *A long life's work*. London: Macmillan and Co.

Geikie, J. 1900. A white-hot liquid Earth and geological time. *Scottish Geographical Magazine* 16: 60–67.

Giesecke, C.L. 1825a. A descriptive catalogue of the different mineral substances found in the vicinity of the Giant's Causeway. *Dublin Philosophical Journal and Scientific Review* 1: 100–108.

Giesecke, C.L. 1825b. On the geological situation of Beryl, discovered in the County of Down. *Dublin Philosophical Journal and Scientific Review* 1: 186–188.

Giesecke, C.L. 1826. *Account of a mineralogical excursion to the County of Donegal*. Dublin: R. Graisberry

Giesecke, C.L. 1828. *Second account of a mineralogical excursion to the Counties of Donegal, Mayo and Galway*. Dublin: R. Graisberry

Giesecke, C.L. 1829. *Account of a mineralogical excursion to the County of Antrim*. Dublin: R. Graisberry

Giesecke, C.L. 1832a. *An Irish mineralogy; or, a list of simple mineral substances hitherto discovered in Ireland*. Dublin: R. Graisberry.

Giesecke, C.L. 1832b. *A descriptive catalogue of a new collection of minerals in the museum of the Royal Dublin Society to which is added an Irish Mineralogy*. Dublin: R. Graisberry.

Gill, M.A.V. and Knell, S.J. 1988. Tunbridge Wells Museum: geology and George Abbott (1844–1925). *The Geological Curator* 5: 3–16.

Gill, W.D. 1956. The Iveagh Geological Laboratory, Trinity College, Dublin. *Nature* 177: 1056–1057.

Gillispie, C.C. 1951. *Genesis and geology*. Cambridge, MA: Harvard University Press.

Glennie, K.W., Brooks, J. and Brooks, J.R.V. 1987. Hydrocarbon exploration and geological history of North West Europe In Brooks, J. and Glennie, K.W. (eds.) *Petroleum geology of North West Europe*. Volume 1. London: Graham and Trotman, 1–10.

Going, R.J. 1984. *A pioneer of the Madras Railways 1857–1875. Diaries of Thomas Hardinge Going. Unpublished memoir.* Copy in the Institute of Engineers, Dublin.

Gough, G.C. 1907. A case of metamorphism of Chalk. *Geological Magazine* 4: 145–148.

Graves, H., Hillier, G.L. and Susan (Harris), Countess of Malmesbury. 1898. *Cycling, part 1-general and mechanical; part 2-cycle racing: part 3-cycling for women*. London: Lawrence and Bullen.

Graydon, G. 1794. On the fish enclosed in stone of Monte Bolca. *Transactions of the Royal Irish Academy* 5: 281–317.

Graydon G. 1802. An account of the straw-hat manufactory, at Newport Pratt, County of Mayo. *Transactions of the Dublin Society* 3: 66.

Graydon, G. 1816. On the Dykes of Monte Somma, in Italy, extracted from a series of letters addressed to the late Lord Charlemont. *Transactions of the Geological Society of London* 3: 233–235.

Green, C.P. 1989. Excursions in the past: a review of the Field Meeting Reports in the first one hundred volumes of the Proceedings. *Proceedings of the Geologists' Association* 100: 17–29.

Greene, M.T. 1982. *Geology in the nineteenth century: changing views of a changing world*. Ithaca, New York and London: Cornell University Press.

Greenwood, T. 1888. *Museums and art galleries*. London: Simkin, Marshall and Co.

Greenwood, T. 1982. Listing of museums. *The Geological Curator* 3: 326–332

Gregory, J.W. 1921. The age of the Earth. *Nature* 108: 283–284.

Griffith, R.J. 1814. *Geological and mining report on the Leinster coal district*. Dublin: Graisberry and Campbell.

Griffith, R.J. 1838. Geological map of Ireland. *Atlas to accompany 2d report of the Railway Commissioners, Ireland*. Dublin.

Griffith, R.J. 1842. *Notice respecting the fossils of the Mountain Limestone of Ireland, as compared with those of Great Britain, and also with the Devonian System*. Dublin: Graisberry and Gill.

Griffith, R.J. 1854. Response to the president. *Proceedings of the Geological Society of London* 10: xx–xxi.

Griffith, R.J. 1860. The localities of the Irish Carboniferous fossils, arranged according to the stratigraphical subdivisions of the Carboniferous System adopted in the Geological Map of Ireland, with the Irish mining localities. *Journal of the Geological Society of Dublin* 9: 21–155.

Griffith, R.J. 1862. List of the fossil localities. In M'Coy, F. *A synopsis of the characters of the Carboniferous limestone fossils of Ireland*. 2nd edition. London: Williams & Norgate.

Grison, E., Goupil, M. and Bret, P. 1994. *Scientific correspondence during the chemical revolution*. Berkeley: Office for the History of Science and Technology, University of California at Berkeley.

Grout, A. 1996. Possessing the Earth: geological collections, information and education in India, 1800–1850. In Crook, N. (ed.) *The transmission of knowledge in South Asia*. Delhi: Oxford University Press, 245–279.

Guettard, J.-E. 1746. Mémoire et carte minéralogique sur la nature et la situation des terreins qui traversent la France & l'Angleterre. *Mémoires de l'Académie Royale des Sciences* 1746: 363–392.

Hackney, P., James, K.W. and Ross, H.C.G. 1983. *A list of photographs in the R.J. Welch collection in the Ulster Museum, Volume 2: Botany, Geology, and Zoology*. Belfast: Ulster Museum.

Hall, J. 1805. Experiments on Whinstone and Lava. *Transactions of the Royal Society of Edinburgh* 5: 43–75.

Halley, E. 1715. A short account of the cause of salti-ness of the oceans, and of the several lakes that emit no rivers; with a proposal by help thereof, to discover the age of the world. *Philosophical Transactions* 29(344): 296–300.

Hamilton, C.W. 1846. Presidential address. *Journal of the Geological Society of Dublin* 3: 168–178.

Hamilton H. 1794. On the power of fixed caustic alkaline salts to preserve the flesh of animals from putrefaction. In a letter to the Reverend George Graydon. *Transactions of the Royal Irish Academy* 5: 319–324.

Hamilton, W. 1772. *Observations on mount Vesuvius, mount Etna, and other volcanos*. London: T. Cadell.

Hamilton, W. 1786. *Letters concerning the Northern Coast of the County of Antrim; respecting the antiquities, manners and customs of that country, together with the natural history of the Basaltes, and its attendant fossils in the northern counties of Ireland; illustrated by an accurate map, and engravings; in two parts*. London: G. Robinson and P. Elmsly.

Hancock, E.G. and Pettitt, C.W. (eds.) 1981. *Register of natural science collections in North West England*. Manchester: Manchester Museum.

Häntzschel, W. 1962. Trace fossils and problematica. In Moore, R.C. (ed.) *Treatise on Invertebrate Paleontology Part W. Miscellanea*. Boulder, CO and Lawrence, KS: Geological Society of America and University of Kansas Press.

Häntzschel, W. 1975. Trace fossils and problematica (revised and enlarged). In Moore, R.C. (ed.) *Treatise on Invertebrate Paleontology Part W. Miscellanea, Supplement 1*. Boulder, CO and Lawrence, KS: Geological Society of America and University of Kansas Press.

Harker, A. 1914. Some remarks on geology in relation to the exact sciences, with an excursus on geological time. *Proceedings of the Yorkshire Geological Society* 19: 1–13.

Harper, D.A.T. (ed.) 1988a. *William King D.Sc.: a palaeontological tribute*. Galway: Officina typographica, Galway University Press.

Harper, D.A.T. 1988b. The King of Queen's College: William King D.Sc., First professor of geology at Galway. In Harper, D.A.T. (ed.) *William King D.Sc. – a palaeontological tribute*. Galway: Officina typographica, Galway University Press.

Harper, D.A.T. and Parkes, M.A. 1996. Geological Survey donations to the Geological Museum in Queens College Galway: 19th century inter-institutional collaboration in Ireland. *The Geological Curator* 6: 233–236.

Hart, J. 1825. *A description of the skeleton of the fossil deer of Ireland, Cervus megaceros, drawn up at the instance of the committee of Natural Philosophy of the Royal Dublin Society*. Dublin: Graisberry [2nd edition 1830].

Hartley, M.M., Norris, A., Pettitt, C.W., Riley, T.H. and Stier, M.A. (eds.) 1987. *Register of natural science collections in Yorkshire and Humberside*. Leeds: Area Museum and Art Gallery Service for Yorkshire and Humberside.

Haughton, S. 1865. *Manual of geology*. (First edition). London: Longman, Green, Longman, Roberts and Green.

Haughton, S. 1871a. *Manual of geology*. (Third edition). London: Longman, Green, Reader and Dyer.

Haughton, S. 1871b. Additional notice of the zeolites of western India. *Journal of the Royal* Geological *Society of Ireland* 2: 112–114.

Haughton, S. 1878. A geological proof that changes in climate in past times were not due to changes in position of the Pole; with an attempt to assign a minor limit to the duration of geological time. *Nature* 18: 266–268.

Hegarty, S. 2017. Du Noyer in the ordnance survey – illustrating Ireland's geology. In Murray, P. (ed.) *Stones, slabs and seascapes: George Victor Du Noyer's images of Ireland*. Cork: Crawford Art Gallery, 21–25.

Hennig, J. 1951. Irish minerals in the Vienna collections. *Irish Naturalists' Journal* 10: 195–196.

Henry, J. 2014. The first geological map of a country. William Smith's 'A Delineation of the Strata of England and Wales'. *International Map Collectors Society Journal* 139: 17–27, 29–30.

[Herries] Davies, G.L. 1962. Joseph Beete Jukes and the rivers of southern Ireland a century's retrospective. *Irish Geography* 4: 221–233.

[Herries] Davies, G.L. 1965. The Geological Society of Dublin and the Royal Geological Society of Ireland 1831–1890. *Hermathena* 100: 66–76.

[Herries] Davies, G.L. 1966. "Basket of Eggs" topography. *Journal of Glaciology* 6: 466.

[Herries] Davies, G.L. 1969a. The University of Dublin and two pioneers of English geology: William Smith and John Phillips. *Hermathena* 109: 24–36.

[Herries] Davies, G.L. 1969b. *The earth in decay*. New York: Elsevier.

[Herries] Davies, G.L. 1970a. The palaeontological collection of Lord Cole, Third Earl of Enniskillen (1807–1886), at Florence Court, Co. Fermanagh. *Irish Naturalists' Journal* 16: 379–381.

[Herries] Davies, G.L. 1970b. Early discoverers 28: Richard Prior's 1699 description of an Irish Esker. *Journal of Glaciology* 9: 147–148.

[Herries] Davies, G.L. 1973. The oldest surviving geological map of Ireland. *Irish Naturalists' Journal* 17: 397–399.

[Herries] Davies, G.L. 1974. First official geological survey in the British Isles. *Nature* 249: 407.

[Herries] Davies, G.L. 1977a. Notes on the various issues of Sir Richard Griffith's quarter-inch geological map of Ireland, 1839–1855. *Imago Mundi* 29: 35–44.

[Herries] Davies, G.L. 1977b. The making of Irish Geography II: Grenville Arthur James Cole (1859–1924). *Irish Geography* 10: 90–94.

Herries Davies, G.L. 1978a. Geology in Ireland before 1812: a bibliographical outline. *The Western Naturalist* 7: 79–99.

Herries Davies, G.L. 1978b. The earth sciences in Irish serial publications 1787–1977. *Journal of Earth Sciences Royal Dublin Society* 1: 1–23.

Herries Davies, G.L. 1979. The making of Irish Geography IV: the physico-historical society of Ireland 1744–1752. *Irish Geography* 12: 92–98.

Herries Davies, G.L. 1980. Richard Griffith – his life and character. In Herries Davies, G.L. and Mollan, R.C. (eds.) 1980. *Richard Griffith: 1784–1878.* Dublin: Royal Dublin Society, 1–31.

Herries Davies, G.L. 1981. The Neptunian and Plutonic theories. In Smith, D.G. (ed.) *Cambridge encyclopaedia of earth sciences.* Cambridge: Cambridge University Press, 12–23.

Herries Davies, G.L. 1983. *Sheets of many colours – the mapping of Ireland's rocks 1750–1890.* Dublin: Royal Dublin Society.

Herries Davies, G.L. 1985a. Astronomy, geology, meteorology. In O'Raifeartaigh, T. (ed.) *The Royal Irish Academy: a bicentennial history 1785–1985.* Dublin: Royal Irish Academy, 247–274.

Herries Davies, G.L. 1985b. Cappoquin, County Waterford. In Mollan, R.C., Davis, W.J. and Finucane, B. (eds.) *Some people and places in Irish science and technology.* Dublin: Royal Irish Academy, 92–93.

Herries Davies, G.L. 1985c. Irish thought in science. In Kearney, R. (ed.) *The Irish mind: exploring intellectual traditions.* Dublin: Wolfhound Press, 294–310.

Herries Davies, G.L. 1990a. Kiltorcan, County Kilkenny. In Mollan, R.C., Davis, W.J. and Finucane, B. (eds.) *More people and places in Irish science and technology.* Dublin: Royal Irish Academy, 86–87.

Herries Davies, G.L. 1990b. Lough Doon, County Kerry. In Mollan, R.C., Davis, W.J. and Finucane, B. (eds.) *More people and places in Irish science and technology.* Dublin: Royal Irish Academy, 88–89.

Herries Davies, G.L. 1995. *North from the Hook: 150 years of the geological survey of Ireland.* Dublin: Geological Survey of Ireland.

Herries Davies, G.L. 2007. *Whatever is under the Earth. The geological society of London 1807 to 2007.* London: Geological Society.

Herries Davies, G.L. and Mollan, R.C. (eds.) 1980. *Richard Griffith: 1784–1878.* Dublin: Royal Dublin Society.

Higgs, B. and Wyse Jackson, P.N. 2007. The role of women in the history of geological studies in Ireland. In Burek, C.V. and Higgs, B. (eds.) *The role of women in the history of geology.* Geological Society, London, Special Publications 281: 137–153.

Holland, C.H. (ed.) 1981. *A geology of Ireland.* Edinburgh: Scottish Academic Press.

Holland, C.H. 1988. The fossiliferous Silurian rocks of the Dunquin inlier, Dingle Peninsula, Co. Kerry, Ireland. *Transactions of the Royal Society of Edinburgh* 79: 347–360.

Holland, C.H. (ed.) 2001. *The geology of Ireland.* Edinburgh: Dunedin Academic Press.

Holland, C.H. and Sanders, I.S. (eds.) 2009. *The geology of Ireland.* 2nd edition. Edinburgh: Dunedin Academic Press.

Holmes, A. 1911. The association of lead with uranium in rock-minerals and its application to the measurement of geological time. *Proceedings of the Royal Society* A85: 248–256.

Holmes, A. 1913. *The age of the earth.* London and New York: Harper & Brothers.

Holmes, A. 1926. Estimates of geological time, with special reference to thorium minerals and uranium haloes. *Philosophical Magazine*, Series 7, 1: 1055–1074.

Holmes, A. 1947. The construction of a geological time-scale. *Transactions of the Geological Society of Glasgow* 21: 117–152.

Holmes, A. and Lawson, R.W. 1927. Factors involved in the calculation of radioactive minerals. *American Journal of Science* 13: 327–244.

Hopkins, W. 1852. Award of the Wollaston medal and donation fund. *Proceedings of the Geological Society of London* for 1852, xix–xx.

Hoppen, K.T. 1970. *The common scientist in the seventeenth century: a study of the Dublin Philosophical Society 1683–1708*. London: Routledge and Kegan Paul.

Horner, L. 1862. Dr W.H. Fitton. *Proceedings of the Geological Society of London for 1862*, xxx–xxxiv.

Howarth, O.R.J. 1931. *The British Association for the Advancement of Science: a retrospect 1831–1931*. London: British Association for the Advancement of Science.

Hudson, R.G.S., Clarke, M.J. & Sevastopulo, G.D. 1966. A detailed account of the fauna and age of a Waulsortian knoll reef limestone and associated shales, Feltrim, Co. Dublin, Ireland. *Scientific Transactions of the Royal Dublin Society* 2A: 251–272.

Hull, E. 1880. On the geological relations of the rock of the south of Ireland with those of North Devon and other British and Continental districts. *Quarterly Journal of the Geological Society of London* 36: 255–274.

Huxley, T.H. 1869. Anniversary address of the President. *Proceedings of the Geological Society* 25: 397–426.

Huxley, T.H. and Wright, E.P. 1871. On a collection of fossil vertebrata from the Jarrow Colliery, County of Kilkenny, Ireland. *Transactions of the Royal Irish Academy* 24, 351–370.

Inglis, H.R.G. 1896. *The 'Contour' road book of Scotland*. Edinburgh: Gall and Inglis.

Inglis, H.R.G. 1897. *The 'Contour' road book of England*. Edinburgh: Gall and Inglis.

Inglis, H.R.G. 1908. *The 'Contour' road book of Ireland*. Edinburgh: Gall and Inglis.

Ingram, T.D. 1887. *A history of the legislative union of Great Britain and Ireland*. London: Macmillan and Co.

Jackson, J.S. 1966. Robert George Spencer Hudson 1895–1965. *Scientific Proceedings of the Royal Dublin Society* Series A, 2: i–ix.

James, K.W. 1986. *'Damned Nonsense!' – The geological career of the third earl of Enniskillen*. Belfast: Ulster Museum.

Johnston, E.M. 1957. State of the Irish House of Commons in 1791. *Proceedings of the Royal Irish Academy* 59C: 1–56.

Joly, J. 1885. Notes on the minerals of the Dublin and Wicklow Granite.—1. Beryl and Iolite of Glencullen. *Scientific Proceedings of the Royal Dublin Society* 5: 48–72.

Joly, J. 1886a. On the permanency of frost-marks, and a possible connexion therewith with *Oldhamia radiata* and *O. antiqua*. *Scientific Proceedings of the Royal Dublin Society* 5: 156–158.

Joly, J. 1886b. On the occurrence of Harmotome at Glendalough, Co. Wicklow. *Scientific Proceedings of the Royal Dublin Society* 5: 165–168.

Joly, J. 1886c. On a peculiarity in the nature of the impressions of *Oldhamia antiqua* and *O. radiata*. *Scientific Proceedings of the Royal Dublin Society* 5: 445–447.

Joly, J. 1886d. On the specific heats of minerals. *Proceedings of the Royal Society* 41: 250–274.

Joly, J. 1896. On a method of photography in natural colours. *Scientific Transactions of the Royal Dublin Society* 6: 127–138.

Joly, J. 1897. On the origin of the canals of Mars. *Scientific Transactions of the Royal Dublin Society* 6: 249–268.

Joly, J. 1898. A shag's meal. *Nature* 59, 125.

Joly, J. 1899. An estimate of the geological age of the Earth. *Scientific Transactions of the Royal Dublin Society* 7: 23–66.

Joly, J. 1900a. A fractionating rain-gauge. *Scientific Proceedings of the Royal Dublin Society* 9: 283–288.
Joly, J. 1900b. Geological age of the Earth. *Geological Magazine*, New Series, Decade 4, 7: 220–225 [Reprinted in *Nature* 62: 235–237].
Joly, J. 1900c. On geological age of the earth. *Report of the British Association for the Advancement of Science (Bradford, 1899)*. London: John Murray, 369–379.
Joly, J. 1900d. Some experiments on denudation in fresh and salt water. *Report of the British Association for the Advancement of Science (Bradford, 1899)*. London: John Murray, 731–732. [Also published in the proceedings of the 8th International Geological Congress, Paris 1901, and enlarged in 1902 in the *Proceedings of the Royal Irish Academy* 24A: 21–32].
Joly, J. 1901a. An estimate of the geological age of the earth. *Annual Report of the Smithsonian Institution (for 1899)*, 247–288.
Joly, J. 1901b. The circulation of salt and geological time. *Chemical News* 83: 301–303.
Joly, J. 1901c. The circulation of salt and geological time. *Geological Magazine* New Series, Decade 4, 8: 354–350.
Joly, J. 1903. Radium and the geological age of the Earth. *Nature* 68, 526.
Joly, J. 1907a. Pleochroic halos. *Philosophical Magazine*, Series 6, 13: 381–383.
Joly, J. 1907b. Distribution of radium in the rocks of the Simplon Tunnel. *Nature* 76: 485.
Joly, J. 1908a. The radioactivity of sea water. *Scientific Proceedings of the Royal Dublin Society* 11: 253–256 [Reprinted in part, and expanded, in *Philosophical Magazine* Series 6, 15: 385–393].
Joly, J. 1908b. On the radium-content of deep-sea sediments. *Scientific Proceedings of the Royal Dublin Society* 11: 288–294 [reprinted, with additions, in *Philosophical Magazine*, Series 6, 16: 190–197].
Joly, J. 1909a. On the distribution of thorium in the Earth's surface materials. *Philosophical Magazine*, Series 6, 18: 140–145.
Joly, J. 1909b. On the radium-content of sea water. *Philosophical Magazine*, Series 6, 18: 396–407.
Joly, J. 1909c. *Radioactivity and geology*. London: Archibald Constable and Co.
Joly, J. 1910. Pleochroic halos. *Philosophical Magazine*, Series 6, 19: 327–330.
Joly, J. 1911a. The age of the Earth. *Philosophical Magazine*, Series 6, 22: 357–380 [reprinted in the *Annual Report of the Smithsonian Institution (for 1911)*, 271–293 (1912)].
Joly, J. 1911b. Radiant Matter. *Scientific Proceedings of the Royal Dublin Society* 13: 73–87.
Joly, J. 1912. The radioactivity of the rocks of the St Gothard Tunnel. *Philosophical Magazine*, Series 6, 23: 201–211.
Joly, J. 1913. Pleochroic haloes. *Bedrock* 4: 453. [Reprinted in the *Annual Report of the Smithsonian Institution (for 1914)*, 313–327].
Joly, J. 1915. *Birth-time of the world and other scientific essays*. London: T. Fisher Unwin.
Joly, J. 1917. The genesis of pleochroic haloes. *Philosophical Transactions of the Royal Society* 217: 51–79.
Joly, J. 1920. *Reminiscences and anticipations*. London: Fisher Unwin.
Joly, J. 1922a. The age of the Earth. *Nature* 109: 480–485 [Reprinted: The age of the Earth, *Royal Institution Library of Science*, Earth Science, 1971, 3(3): 140–151. Friday Evening Discourses in Physical Sciences held at the Royal Institution 1922].
Joly, J. 1922b. Haloes and Earth-history: a new radioactive element. *Nature* 109: 517–518.

Joly, J. 1923a. The movements of the Earth's surface crust. *Philosophical Magazine*, Series 6, 45: 1167–1188.

Joly, J. 1923b. Pleochroic haloes of various geological ages. *Proceedings of the Royal Society*, Series A, 102: 682–705.

Joly, J. 1925. *Surface history of the Earth* (1st edition). Oxford: Clarendon Press.

Joly, J. 1928. The theory of thermal cycles. *Gerlands Beiträge zur Geophysik* 19: 415–441.

Joly, J. 1930. *Surface history of the Earth* (2nd edition). Oxford: Clarendon Press.

Joly, J. 1931. History of the Irish Radium Institute (1914–1930). *Royal Dublin Society Bicentenary Souvenir 1731–1931*. Dublin: Royal Dublin Society, 23–32.

Joly, J. and Fletcher, A.L. 1910. Pleochroic halos. *Philosophical Magazine*, Series 6, 19: 631–648.

Joly, J. and Poole, J.H.J. 1927. On the nature and origin of the Earth's surface structure. *Philosophical Magazine*, Series 7, 3: 327–330.

Joly, J. and Rutherford, E. 1913. The age of pleochroic haloes. *Philosophical Magazine*, Series 6, 25: 644–657.

Jørgensen, G. 1983. Karl Ludwig Giesecke-fra Mozartsanger til Grønlandspioner. *Tidsskriftet Grønland* 6: 183–188.

Jørgensen, G. 1996. Charles Lewis Giesecke, professor of mineralogy in Dublin a fascinating character in the geological history of the Faeroe Islands and Greenland. *Irish Journal of Earth Sciences* 15: 155–160.

Jukes, J.B. 1862. Student's Manual of Geology. Edinburgh: A. & C. Black.

Jukes, J.B. 1866. On the Carboniferous slate (or Devonian rocks) and the Old Red Sandstone of south Ireland and North Devon. *Quarterly Journal of the Geological Society of London* 22: 320–371.

Keane, E., Phair, P.B., and Sadleir, T.U. 1982. *King's Inns Admission Papers 1607–1867*. Dublin: Irish Manuscripts Commission.

Kelham, B.B. 1967. The Royal College of Science (1867–1926). *Studies* 56: 297–309.

Kelly, J. 1855. On locations of fossils of the Carboniferous Limestone of Ireland. *Journal of the Geological Society of Dublin* 7: 1–62.

Kelly, P. 1992. From Molyneux to Berkeley: The Dublin Philosophical Society and its Legacy. In Scott, D.S. (ed.) *Treasures of the mind: a Trinity College Dublin quatercentenary exhibition*. London: Sotheby's, 23–31.

Kerr-Lawson, D.E. 1927. Pleochroic haloes in biotite from near Murray Bay, Province of Quebec. *University of Toronto Studies, Geological Series* 24: 54–70.

Kinahan, G.H. and Baily, W.H. 1880. Report on the rocks of the Fintona and Curlew Mountain Districts. *Proceedings of the Royal Irish Academy* 3: 475–500.

King, W. 1850. *A monograph on the Permian fossils of England*. London: Palaeontographical Society.

King, W.B.R. 1955. The geological history of the English Channel. *Quarterly Journal of the Geological Society of London* 110: 77–101.

Kirwan, R. 1793. Examination of supposed igneous origin of stony substances. *Transactions of the Royal Irish Academy* 5: 51–81.

Kirwan, R. 1794. *Elements of mineralogy*. 2nd Edition, Volume 1. London: J. Nichols.

Kirwan, R. 1799. *Geological essays*. London: D. Bremner.

Knauth, L.P. 1998a. Salinity history of the Earth's early ocean. *Nature* 395: 554–555.

Knauth, L.P. 1998b. Salinity, temperature, and oxygen history of the Precambrian ocean. *Abstracts with Programs – Geological Society of America* 30(7): 272.

Knell, S.J. 1996. The roller-coaster of museum geology. In Pearce, S. (ed.) *Exploring science in museums*. London and New Jersey: Athlone Press, 29–56.

Knell, S.J. 2000. *The culture of English geology, 1815–1851: a science revealed through its collecting*. Aldershot: Ashgate.

Knox, G. 1802. Observations on Calp (found at Dublin). *Transactions of the Royal Irish Academy* 8: 207–214.

Knox, G. 1822. Experiments and observations on the Newry pitch-stone, and its products, and on the formation of pumice. *Philosophical Transactions of the Royal Society London* 112: 313–338.

Knox, G. 1823. On Bitumen in stones. *Philosophical Transactions of the Royal Society London* 113: 517–528.

Knox, G. 1826. Enumeration of the simple mineral substances found in the vicinity of Dublin. *Philosophical Journal and Scientific Review* 2: 208–213.

Kölbl-Ebert, M. 2007. The geological travels of Charles Lyell, Charlotte Murchison and Roderick Impey Murchison in France and Northern Italy (1828). In Wyse Jackson, P.N. (ed.) *Four centuries of geological travel: the search for knowledge on foot, bicycle, sledge and camel*. Geological Society, London, Special Publications 287: 109–117.

Kumar, D. 1995. *Science and the Raj*. Delhi: Oxford University Press.

Lane, A.C. 1925. Measurement of geological age by atomic disintegration. *Proceedings – Lake Superior Mining Institute, Annual Meeting* 24(45): 106–116.

Lane, A.C. 1929. The Earth's age by sodium accumulation. *American Journal of Science* Series 5, 17: 342–346.

Laudan, R. 1987. *From mineralogy to geology: the foundations of a science 1650–1830*. Chicago and London: University of Chicago Press.

Lee, J. 1980. Richard Griffith's land valuation as a basis for farm taxation. In Herries Davies, G.L. and Mollan, R.C. (eds.) *Richard Griffith: 1784–1878*. Dublin: Royal Dublin Society, 71–101.

Lemon, K., Wyse Jackson, P.N., Cooper, M. and Dempster, M. 2022. The Paleocene volcanic rocks of the Giant's Causeway and Causeway coast, United Kingdom. In Hilario, A. (ed.) *The first 100 IUGS Geological Heritage Sites*. Zumaia: International Union of Geological Sciences, 30–31, 270–271.

Leviton, A. and Aldrich, M.L. 2000. India: a case study of natural history in a colonial setting. *Memoirs of the California Academy of Sciences* 25: 51–80.

Lewis, C. 2000. *The dating game: one man's search for the age of the Earth*. Cambridge: Cambridge University Press.

Lewis, C.L.E. 2001. Arthur Holmes' vision of a geological timescale. In Lewis, C.L.E. and Knell, S.J. (eds.) *The age of the Earth: from 4004 BC to AD 2002*. Geological Society, London, Special Publications 190: 121–138.

Lewis, C.L.E. and Knell, S.J. (eds.) 2009. *The making of the Geological Society of London*. Geological Society, London, Special Publications 317.

Livesey, S.J. 1998. De viris illustribus et mediocribus: a biographical database of Franciscan commentators on Aristotle and Peter Lombard's Sentences. *Franciscan Studies* 56: 203–237.

Long, A. 1921a. Notes on the career of Sir Charles L M Giesecke. *Journal of the Irish Memorials Association* 11: 35–38.

Long, T. 1921b. Dublin St George's Church. *Journal of the Irish Memorials Association* 11: 32–34.

M'Coy, F. 1841. *A catalogue of the museum of the Geological Society of Dublin*. Dublin: Hodges and Smith.

M'Coy, F. 1844. *A synopsis of the characters of the Carboniferous limestone fossils of Ireland*. Dublin: Dublin University Press.

M'Coy, F. 1846. *A synopsis of the characters of the Silurian fossils of Ireland*. Dublin: Dublin University Press.

M'Coy, F. 1848. On some new fossil fish of the Carboniferous Period. *Annals and Magazine of Natural History* 2: 115–133.

M'Coy, F. 1854. *Contributions to British palaeontology*. Cambridge: Macmillan and Co.

M'Coy, F. 1862a. *A synopsis of the characters of the Carboniferous limestone fossils of Ireland*. 2nd edition. London: Williams & Norgate.

M'Coy, F. 1862b. *A synopsis of the characters of the Silurian fossils of Ireland*. 2nd edition. London: Williams & Norgate.

MacArthur, C.W.P. 1987. Mineralogical and geological travellers in Donegal 1787–1812. *Donegal Annual* 39: 39–57.

MacArthur, C.W.P. 1990. Dr Jean-Francois Berger of Geneva (1779–1833) from the travelling fund to the Wollaston Donation. *Archives of Natural History* 17: 97–119.

Malcolmson, A.P.W. 1978. *John Foster: the politics of the Anglo-Irish ascendancy*. Oxford: Oxford University Press.

Marchant, T.R. and Sevastopulo, G.D. 1980. The Calp of the Dublin district. *Journal of Earth Sciences, Royal Dublin Society* 3: 195–203.

McArdle, P. 2018. *Dissenting spirit: Thomas Weaver, geologist and mining engineer*. Dublin: The Liffey Press.

McCartney, P.J. 1977. *Henry De la Beche: observations on an observer*. Cardiff: Friends of the National Museum of Wales.

McDowell, R.B. and Webb, D.A. 1982. *Trinity College Dublin: an academic history*. Cambridge: Cambridge University Press.

McHenry, A. 1896. Long excursion to Belfast, the coast of Antrim, and the Mourne Mountains. *Proceedings of the Geologists' Association* 14: 200–206.

McKenna, G. 1990. Geological photographs of the British Association for the Advancement Science. *Geology Today* 6: 157–159.

McLaughlin, P.J. 1939–1940. Richard Kirwan. *Studies* 28: 461–474, 593–605; 29: 71–83, 281–300.

McLaughlin, P.J. 1954. The Kirwanian Society (1812). *Studies* 43: 441–450.

McMillan, N.D. 1988. Rev. Samuel Haughton and the Age of the Earth Controversy. In Nudds, J.R., McMillan, N.D., Weaire, D.L. and McKenna Lawlor, S.M.P. (eds.) *Science in Ireland 1800–1930: tradition and reform*. Dublin: Trinity College Dublin, 151–161.

McNairn, W.H. 1919. How old is the world? *Popular Astronomy* 27(2): 1–8.

Medawar, P.B. 1979. *Advice to a young scientist*. New York and London: Harper and Row.

Meenan, J. and Clarke, D. (eds.). 1981. *RDS: the Royal Dublin Society 1731–1981*. Dublin: Gill and Macmillan.

Mellard Reade, T. 1877. President's address. *Proceedings of the Liverpool Geological Society* 3(3): 211–235.

Mellard Reade, T. 1878. The age of the world as viewed by the geologist and the mathematician. *Geological Magazine*, New Series, Decade 2, 5: 145–154.

Mellard Reade, T. 1879. *Chemical denudation in relation to geological time*. London: Daniel Dogue.

Mellard Reade, T. 1893. Measurement of geological time. *Geological Magazine*, New Series, Decade 3, 10: 97–100.

Milner, A.C. 2019. A morphological revision of *Keraterpeton*, the earliest horned nectridean from the Pennsylvanian of England and Ireland. *Earth & Environmental Science Transactions of the Royal Society of Edinburgh* 109: 237–253.

Milner, A.R. 1980. The temnospondyl amphibian *Dendrerpeton* from the Upper Carboniferous of Ireland. *Palaeontology* 23, 125–141.

Mitchell, G.F. 1981. Mineralogy and geology. In Meenan, J. and Clarke, D. (eds.) *The Royal Dublin Society 1731–1981.* Dublin: Gill and Macmillan, 154–169.

Mitchell, G.F. 1990. Anthony Farrington, geomorphologist and quaternary geologist, 1893–19. In Mollan, R.C., Davis, W.J. and Finucane, B. (eds.) *More people and places in Irish science technology.* Dublin: Royal Irish Academy, 64–65.

Mittlefehldt, D.W. 1999. Sodium. In Marshall, C.P. and Fairbridge, R.W. (eds.) *Encyclopedia of geochemistry.* Dordrecht: Kluwer, 577–578.

Mollan, R.C. 1990. Royal Dublin Society. In Mollan, R.C., Davis, W.J. and Finucane, B. (eds.) *More people and places in Irish science and technology*. Dublin: Royal Irish Academy, 92–95.

Molyneux, T. 1694. Some notes upon the foregoing account of the Giants Causeway, serving to further illustrate the same. *Philosophical Transactions of the Royal Society of London* 18: 175–182.

Molyneux, T. 1697. A discourse concerning the large horns frequently found under ground in Ireland, concluding from them that the Great American Deer, call'd a Moose, was Common in that Island. *Philosophical Transactions of the Royal Society of London* 19: 489–512.

Monaghan, N.T. 1992. Geology in the national museum of Ireland. *The Geological Curator* 5: 275–282.

Monaghan, N.T. 1993. Sir Karl Ludwig Metzler-Giesecke (1761–1833), royal minerologist, Greenland explorer and museum curator. In Hoch, E. and Brantsen, A.K. (eds.) *Deciphering the natural world and the role of collections and museums.* Copenhagen: Geologisk Museum, 82–86.

Monaghan, N.T. 2001. Irish palaeontological illustrations of the 19th century. In Rushton, B.S., Hackney, P. and Tyrie, C.R. (eds.) *Biological collections & biodiversity*. Linnean Society Occasional Publications 3: 83–91.

Monaghan, N.T. 2017. The scientific illustrations of George Victor Du Noyer. In Murray, P. (ed.) *Stones, slabs and seascapes: George Victor Du Noyer's images of Ireland.* Cork: Crawford Art Gallery, 27–33.

Monaghan, N.T. 2025. The fossil collections of Sir Richard John Griffith (1784–1878). *Irish Journal of Earth Sciences* 43.

Moore, D.T. 1982. An account of those described rock collections in the British Museum (Natural History) made before 1918; with a provisional catalogue arranged by continent. Bulletin of the British Museum (Natural History) (Historical Series) 10: 141–177.

Moro, A.L. 1740. *De' Crostacei e degli altri Marini Corpi the si truovano su' monti.* Venezia: S. Monti.

Morrell, J. 1990. The British Association in Ireland. In Mollan, R.C., Davis, W.J. and Finucane, B. (eds.) *More people and places in Irish science and technology.* Dublin: Royal Irish Academy, 78–81

Morrell, J. 2001. Genesis and geochronology: the case of John Phillips (1800–1874). In Lewis, C.L.E. and Knell, S.J. (eds.) *The age of the Earth: from 4004 BC to AD 2002.* Geological Society, London, Special Publications 190: 85–90.

Morrell, J. 1994. Perpetual excitement: the heroic age of British geology. *The Geological Curator* 5: 311–317.

Morrell, J. and Thackray, J. 1981. *Gentlemen of science: early years of the British Association for the Advancement of Science*. Oxford: Oxford University Press.

Murnaghan, D. 1985. The Irish Radium Institute, Dublin. In Mollan, R.C., Davis, W.J. and Finucane, B. (eds.) *Some people and places in Irish science and technology*. Dublin: Royal Irish Academy, 102–103.

Murray, J. and Renard, A.F. 1891. *Report on the deep-sea deposits based on the specimens collected during the voyage of the H.M.S. Challenger in the years 1872–1876*. London: Her Majesty's Stationary Office.

Murray, J., Nasheuer, H.P., Seoighe, C., McCormack, G.P., Williams, D.M. and Harper, D.A.T. 2015. The contribution of William King to the early development of Palaeoanthropology. *Irish Journal of Earth Sciences* 33: 1–16.

Murray, P. (ed.) 2017. *Stones, Slabs and Seascapes: George Victor Du Noyer's images of Ireland*. Cork: Crawford Art Gallery

Naylor, D. and Shannon, P.M. 1982. *The geology of offshore Ireland and west Britain*. London: Graham and Trotman.

Nelson, E.C. 2000. A caricature of an Irish naturalist: Revd Dr William Richardson (1740–1820). *Archives of Natural History* 27: 149–151.

Nesbitt, N. 1979. *A museum in Belfast: A history of the Ulster Museum and its predecessors*. Belfast: Ulster Museum.

Newcomb, S. 2012. Richard Kirwan (1733–1812). *Earth Sciences History* 31: 287–314.

Newmann, K. 1993. *Dictionary of Ulster Biography*. Belfast: Institute of Irish Studies, Queen's University of Belfast.

Nimmo, A. 1814. Connemara, Co. Galway; Counties of Kerry and Cork. In: Anon. *The fourth report of the commissioners appointed to enquire into the nature and extent of several bogs in Ireland and the practicability of draining and cultivating them*. London: House of Commons.

Nimmo, A. 1824. *Report of the select committee on the survey and valuation of Ireland*. London: House of Commons.

Nimmo, A. 1825a. On the application of the science of geology to the purposes of practical navigation. *Transactions of the Royal Irish Academy* 14: 39–50.

Nimmo, A. 1825b. On the application of the science of geology to the purposes of practical navigation. *Dublin Philosophical Journal and Scientific Review* 1: 152–161.

Nimmo, A. 1831. *Charts of the harbours and coastline of Ireland*. London: Josiah Neele.

Nudds, J.R. (ed.) 1983. *Upon sweet mountains: a selection of poetry by John Joly, F.R.S.* Dublin: Trinity Closet Press.

Nudds, J.R. 1986. The life and work of John Joly (1857–1933). *Irish Journal of Earth Sciences*, 8: 81–94.

Nudds, J.R. 1988. John Joly; brilliant polymath. In Nudds, J.R., McMillan, N.D., Weaire, D.L. and McKenna Lawlor, S.M.P. (eds.) *Science in Ireland 1800–1930: tradition and reform*. Dublin: Trinity College Dublin, 163–178.

Nudds, J.R. (ed.) 1994. *Directory of British geological museums*. Geological Society, London, Miscellaneous Paper 18.

Ó Gogáin, A. and Wyse Jackson, P.N. 2021. Microcomputed tomography of the holotype of the early tetrapod *Ichthyerpeton bradleyae* (Huxley in Wright and Huxley, 1866) from the Pennsylvanian of Ireland. *Journal of Paleontology* 95: 1048–1060.

Ó Maithú, S. 2003. *Dublin's suburban towns, 1834–1930*. Dublin: Four Courts Press.

O'Conor, R.C. 1949. Origins of the Royal Irish Academy. *Studies*, Sept. 1949: 325–337.

O'Donnell, S. 1984. Irish time. *Eire-Ireland* 19: 131–134.

O'Raifeartaigh, T. (ed.) 1985. *The Royal Irish Academy: a bicentennial history 1785–1985*. Dublin: Royal Irish Academy.

O'Riordan, C.E. 1983. *The Natural History Museum, Dublin*. Dublin: Stationery Office.

Oldroyd, D.R. 1990. *The Highlands controversy: constructing geological knowledge through fieldwork in nineteenth-century Britain*. Chicago: Chicago University Press.

Oreskes, N. 1999. *The rejection of continental drift*. Oxford: Oxford University Press.

Parkes, M.A. 2021. George Victor Du Noyer's large format paintings: nineteenth-century lecture slides. In Clary, R.M., Rosenberg, G.D. and Evans, D.C. (eds.) *The evolution of paleontological art*. Geological Society of America Memoir 218: 101–108.

Parkes, M.A., Coffey, P. and Connaughton, P. 2000. The printing wood block collection of the Geological Survey of Ireland. *The Geological Curator* 7: 149–156.

Parkes, M.A. and Owen, A.W. 1994. Anomalous records of trinucleid trilobites in Irish Silurian rocks. *Irish Journal of Earth Sciences* 13: 59–63.

Parkes, M.A. and Wyse Jackson, P.N. 1998. A survey of the state and status of geological collections in museums and private collections in the Republic of Ireland. *The Geological Curator* 6: 377–388.

Parkin, J. 1976. Silurian rocks of the Bull's Head, Annascaul and Derrymore Glen inliers, Co. Kerry. *Proceedings of the Royal Irish Academy* 76B: 577–606.

Pemberton, S.G. 2010. History of ichnology: Earth ichnology poems and their poets. *Ichnos*, 17: 264–270.

Penny, L.F. 1966. Early discoverers XXIV: George William Lamplugh (1859–1926). *Journal of Glaciology* 6: 307–309.

Petersen, O.V. 1993. The long tradition of Greenland mineral collecting, with special emphasis on the type specimens in the mineral collection of the Geological Museum. In Hoch, E. and Brantsen, A.K. (eds.) *Deciphering the natural world and the role of collections and museums*. Copenhagen: Geologisk Museum, 89–91.

Petersen, O.V. and Secher, K. 1993. The minerals of Greenland – introduction Giesecke. *Mineralogical Record* 24: 7–9.

Pettigrew, T.H. 1979. William King (1808–1886) – a biographical note. *Newsletter of the Geological Curators' Group* 2: 327–329.

Pettit, S.F. 1976. The Royal Cork Institution: a reflection of the cultural life of a city. *Journal of the Cork Historical and Archaeological Society* 81: 70–90.

Phillips, J. 1844. *Memoirs of William Smith, LL.D., author of the "Map of the Strata of England and Wales"*. London: John Murray.

Phillips, J. 1860. *Life on the Earth: its origin and succession*. Cambridge: Macmillan.

Pini, E. 1792. Memoria Geologica sulle Rivoluzioni del Globo Terrestre prodotte dall'azione delle acque. Parte seconda. *Memorie di Matematica e Fisica della Society Italiana dei XL* 6: 389–500.

Playfair, J. 1802. *Illustrations of the Huttonian Theory of the Earth*. Edinburgh: Cadell and Davies.

Pococke, R. 1748a. An account of the Giants Causeway in Ireland. *Philosophical Transactions* 45: 124–127.

Pococke, R. 1748b. A father account of the Giants Causeway in Ireland, by the Rev. Richard Pococke LL.D. Archdeacon of Dublin, and F.R.S. *Philosophical Transactions* 48: 226–237.

Pococke, R. 1753. A letter upon the same subject from the Rev. Richard Pococke LL.D. Archdeacon of Dublin, to the Rev. Tho. Birch D.D. secr. R.S. *Philosophical Transactions* 48: 238.

Pomian, K. 1986. Collezionisti d'arte e di curiosità naturali. In Arnaldi, G. and Pastore Stocchi, M. (eds.) *Storia della cultura veneta. 5.II Il Settecento.* Vicenza: Neri Pozza, 1–70.

Porter, R. 1977. *The making of geology: earth science in Britain 1660–1815.* Cambridge: Cambridge University Press.

Portlock, J.E. 1843. *Report on the geology of the County of Londonderry, and parts of Tyrone and Fermanagh.* Dublin: A. Milliken, Hodges and Smyth, and Longman, Brown, Green and Longmans.

Portlock, J.E. 1869. *Memoir of the life of Major-General Colby.* London: Seeley, Jackson and Halliday.

Poulton, E.B. 1897. A naturalist's contribution to the discussion upon the age of the Earth. *Report of the sixty-sixth meeting of the British Association for the Advancement of Science (Liverpool, 1896).* London: John Murray, 808–828.

Praeger, R.LL. 1941. *A populous solitude.* Dublin: Hodges Figgis.

Praeger, R.L. 1949. *Some Irish naturalists: a biographical notebook.* Dundalk: Dundalgan

Prestwich, J. 1855–1857. On the correlation of the Eocene Tertiaries of England, France and Belgium. Part 1 (1855), Part 2 (1857). *Quarterly Journal of the Geological Society of London* 11: 206–246; 13: 89–134.

Pringle, J.W. 1827. *Directions for geological and mineralogical observations.* London.

Pyenson, L. 1977. "Who the guys were": prosopography in the history of science. *History of Science* 15: 155–188.

Quin, E.G. and Freeman, T.W. 1947. Some Irish topographical terms. *Irish Geography* 1: 85–89.

Reese, R.L., Everett, S.M. and Craun, E.D. 1981. The chronology of Archbishop James Ussher. *Sky and Telescope*, 62: 404–405.

Reilly, J.C. 2000. *Richard Griffith and his valuations of Ireland.* Baltimore: Genealogical Publishing Company.

Reynolds, D. 1928. The petrography of the Triassic sandstone of north-east Ireland. *Geological Magazine* 65: 448–473.

Reynolds, D. 1931. The dykes of the Ards Peninsula. *Geological Magazine* 68: 97–111.

Reynolds, D. 1934. The eastern end of the Newry Igneous Complex. *Quarterly Journal of the Geological Society of London* 90: 585–636.

Reynolds, D. 1937a. Contact phenomena indicating a Tertiary age for the gabbros of the Slieve Gullion District. *Proceedings of the Geologists' Association* 48: 247–275.

Reynolds, D. 1937b. The two monzonitic series of the Newry Complex. *Geological Magazine* 73: 337–364.

Reynolds, D. 1942. The Albite-Schists of Antrim and their petrogenetic relationship to Caledonian orogenesis. *Proceedings of the Royal Irish Academy* 48B: 43–66.

Reynolds, D. 1943a. Granitization of hornfelsed diorite in the Newry granodiorite of Goraghwood Quarry, Co. Armagh. *Proceedings of the Royal Irish Academy* 48B: 231–267.

Reynolds, D. 1943b. The south-western end of the Newry Igneous Complex. A contribution towards the petrogenesis of the granodiorites. *Quarterly Journal of the Geological Society of London* 99: 205–246.

Reynolds, D. 1946. The sequence of geochemical changes leading to granitisation. *Quarterly Journal of the Geological Society of London* 102: 390–447.

Reynolds, D. 1951. The geology of Slieve Gullion, Foughill, and Carrickcarnon; an actualistic interpretation of a Tertiary gabbro-granophyre complex. *Transactions of the Royal Society of Edinburgh* 62: 85–142.

Reynolds, D. 1954. Fluidization as a geologic process, and its bearing on the problem of intrusive granites. *American Journal of Science* 252: 577–614.

Richardson, W. 1802. On basalts. *Philosophical Magazine* 13: 129–136.

Richardson, W. 1803a. Account of the whynn dykes in the neighbourhood of the Giant's Causeway, Ballycastle, and Belfast. *Transactions of the Royal Irish Academy* 9: 21–43.

Richardson, W. 1803b. Inquiry into the consisting of Dr. Hutton's theory of the Earth with the arrangements of the strata and other phenomena on the basaltic coast of Antrim. *Transactions of the Royal Irish Academy* 9: 429–487.

Richardson, W. 1805. Remarks on the basalts of the coast of Antrim. *Transactions of the Royal Society of Edinburgh* 5: 15–20.

Richardson, W. 1806a. On the Volcanic Theory. Part 3. Examination of the facts and opinions, given by different advocates for the volcanic origin of basalt, who followed Mr. Desmarest; to wit, Mr. Faujas de St. Fond, Mr. Dolomieu, Mr. Whitehurst, Bishop Troll, Abbe Spalanzani, and Dr. Hamilton. *Transactions of the Royal Irish Academy* 10: 63–86.

Richardson, W. 1806b. On the Volcanic Theory. Part 3. Arguments against the volcanic origin of Basalt, derived from its arrangement in the County of Antrim, and from other facts observed in that Country. *Transactions of the Royal Irish Academy* 10: 87–107.

Richardson, W. 1808. A letter on the alterations that have taken place in the structure of rocks on the surface of the basaltic country in the counties of Derry and Antrim. *Transactions of the Royal Society* 98: 187–222.

Richardson, W. 1812. On the basaltic productions, &c, of the county of Antrim, Appendix II. In Dubordieu, J. (ed.) *Statistical survey of the County of Antrim with observations on the means of improvement, drawn up by the order and for the consideration of the Dublin Society.* Dublin: Dublin Society.

Richter, F.M. 1986. Kelvin and the age of the Earth. *Journal of Geology* 95: 395–401.

Rickards, R.B., Burns, V. and Archer J.B. 1973. The Silurian sequence at Balbriggan, Co. Dublin. *Proceedings of the Royal Irish Academy* 73B: 303–316.

Robinson, E. 1990. 'Clarion o'er the dreaming earth': a personal review of the G.A. Circular since 1858. *Proceedings of the Geologists' Association* 101: 101–117.

Rose, E.P.F. 1996. Geologists and the army in nineteenth century Britain: a scientific and educational symbiosis? *Proceedings of the Geologists' Association* 107: 129–142.

Rose, E.P.F. 1999. The military background of John W. Pringle, in 1826 founding Superintendent of the Geological Survey of Ireland. *Irish Journal of Earth Sciences* 17: 61–70.

Rudwick, M.J.S. 1985. *The great Devonian controversy.* Chicago University Press.

Rudwick, M.J.S. 1988. A year in the life of Adam Sedgwick and company, geologists. *Archives of Natural History* 15: 243–268.

Rupke, N.A. 1994. *Richard Owen: Victorian naturalist.* New Haven: Yale University Press.

Rushton, A. 1979. The real M'Coy. *Lethaia* 12: 226.

Rutherford, E. 1905. Present problems in radioactivity. *Popular Science Monthly* 67: 5–34.

Rudzki, M.P. 1901. *Sur l'Age de la Terre. Bulletin de l'Académie des Sciences de Cracovie* 1901, 72–94.

Ryley, J.B. 1899. *The dangers of cycling for women and children.* London: Henry Renshaw.

Sangwan, S. 1994. Reordering the earth: the emergence of geology as a scientific discipline in colonial India. *Indian Economic and Social History Review* 31: 291–310.

Sequeira, S.E.K. 1996. A cochleosaurid amphibian from the Upper Carboniferous of Ireland. *Special Papers in Palaeontology* 52: 65–80.

Scott, E.L. 1973. R. Kirwan. In Gillispie, C.C. (ed.) *Dictionary of scientific biography,* Volume 7. New York: C. Scribner's Sons, 387–390.

Scott, E.L. 1979. *The life and work of Richard Kirwan*. Unpublished PhD thesis, University of London.

[Scouler, J.] 1834. Biographical sketch of Sir Charles Lewis Metzler von Giesecke, late professor of mineralogy to the Royal Dublin Society. *Dublin University Magazine* 3: 161–75, 296–306.

Secord, J.A. 1986. *Controversy in Victorian geology: the Cambrian Silurian dispute*. Princeton: Princeton University Press.

Secord, J.A. 1995. John W. Salter: the rise and fall of a Victorian palaeontological career. In Wheeler, A. and Price, J.H. (eds.) *From Linnaeus to Darwin: commentaries on the history of biology and geology*. Society for the History of Natural History, London Special Publication 3: 61–75.

Sedgwick, A. and M'Coy, F. 1851–1855. *A synopsis of the classification of British Palaeozoic rocks, with a systematic description of the British Palaeozoic fossils in the Geological Museum of the University of Cambridge*. London and Cambridge.

Seymour, H.J. 1944. The centenary of the first geological survey made in Ireland. *Economic Proceedings of the Royal Dublin Society* 3: 227–248.

Shapin, S. and Thackray, J. 1974. Prosopography as a research tool in the history of science: the British scientific community, 1700–1900. *History of Science* 12: 1–28.

Sharpe, T. and McCartney, P.J. 1998. *The papers of H.T. De la Beche (1796–1855) in the National Museum of Wales*. Cardiff: National Museums and Galleries of Wales.

Sheets-Pyenson, S. 1988. How to "grow" a natural history museum: the building of colonial collections, 1850–1900. *Archives of Natural History* 15: 121–147.

Shelton, H.S. 1910. The age of the earth and the saltness of the sea. *Journal of Geology* 18: 190–193.

Shelton, H.S. 1911. Modern theories of geologic time. *The Contemporary Review* 99: 195–209.

Sheppard, T. 1915. William Smith: his maps and memoirs. *Proceedings of the Yorkshire Geological Society* 19: 75–253.

Sherborn, C.D. 1940. *Where is the collection? Account of the various natural history collections which have come under the notice of the compiler*. Cambridge: Cambridge University Press.

Shipley, B. 2001. 'Had Lord Kelvin a right?': John Perry, natural selection and the age of the Earth, 1894–1895. In Lewis, C.L.E. and Knell, S.J. (eds.) *The age of the Earth: from 4004 BC to AD 2002*. Geological Society, London, Special Publications 190: 91–105.

Siegfried, R. and Dott, R.H. Jr. 1980. *Humphry Davy on geology: the 1805 lectures for the general audience*. Madison: University of Wisconsin Press.

Sigwart, J. 2007. The Irish fossil Polyplacophora. *Irish Journal of Earth Sciences* 25: 27–38.

Silotti, A. 1979. *I fossili des Lessini*. (2nd ed.) Verona: Edizioni di Vita Veronese.

Simington, R.C. and Farrington, A. 1949. A forgotten pioneer: Patrick Ganly, geologist, surveyor and civil engineer (1809–1899). *Journal of the Department of Agriculture* 46: 36–50.

Simms, M.J. 2021. Subsidence, not erosion: revisiting the emplacement environment of the Giant's Causeway, Northern Ireland. *Proceedings of the Geologists' Association* 132: 537–548.

Smith, A.S. and Dyke, G.J. 2008. The skull of the giant predatory pliosaur *Rhomaleosaurus cramptoni*: implications for plesiosaur phylogenetics. *Naturwissenschaften* 95: 975–980.

Smith, R.A. 1972. *A social history of the bicycle, its early life and times in America*. New York: American Heritage Press.

Smyth, W.W. 1868. Anniversary address of the President. Award of the Wollaston-donation fund. *Proceedings of the Geological Society* 32: xxvii.

Solinas, G. 1978. *Paleontologia di Bolca, Ronca, S.Giovanni Ilarione. Repertorio bibliografico dal 1552 al 1975*. Verona: Tip. La Grafica.

Sollas, W.J. 1895a. On *Pucksia MacHenryi*, a new fossil from the Cambrian Rocks of Howth. *Scientific Proceedings of the Royal Dublin Society* 8: 297–303.

Sollas, W.J. 1895b. The age of the Earth. *Nature* 51: 558.

Sollas, W.J. 1900. Evolutionary geology. *Report of the British Association for the Advancement of Science (Bradford, 1899)*. London: John Murray, 449–463 [Reprinted 1915 in *The age of the Earth*. London: T. Fisher Unwin].

Sollas, W.J. 1909. Anniversary address of the President: position of geology among the sciences; on time considered in relation to geological events and to the development of the organic world; the rigidity of the Earth and the age of the oceans. *Proceedings of the Geological Society* 65: lxxxi–cxxiv.

Somerfield, A.E. 1985. John Joly geologist, physicist and mineralogist 1857–1933. In Mollan, R.C., Davis, W.J. and Finucane, B. (eds.) *Some people and places in Irish science and technology*. Dublin: Royal Irish Academy, 64–65.

Somerville-Large, P. 1975. *Irish eccentrics*. London: Hamish Hamilton.

Sorbini, L. 1981. *I fossili di Bolca*. 2nd edition. Verona: Tip. La Grafica.

Sorbini, L. 1985. Scipione Maffei e i fossili di Bolca. *La Lessinia – ieri, oggi, domani* 1–2: 97–102.

Spalding, D.A.E. 2007. Two Tyrrells cross the Barren Lands of Canada, 1893. In Wyse Jackson, P.N. (ed.) *Four centuries of geological travel: the search for knowledge on foot, bicycle, sledge and camel*. Geological Society, London, Special Publications 287: 287–296.

Spamer, E.E., Bogan, A.E. and Torrens, H.S. 1989. Recovery of the Etheldred Benett Collection of fossils mostly from Jurassic-Cretaceous strata of Wiltshire, England, analysis of the taxonomic nomenclature of Benett (1831), and notes and figures of type specimens contained in the collection. *Proceedings of the Academy of Natural Sciences of Philadelphia* 141: 115–180.

Spearman, T.D. 1990. Samuel Haughton: mathematician, geologist, anatomist and physiologist – 1821–1897. In Mollan, R.C., Davis, W.J. and Finucane, B. (eds.) *More people and places in Irish science and technology*. Dublin: Royal Irish Academy, 36–37.

Spencer, C. 1884. *The cycle directory: containing an alphabetical list of all Cycling Clubs & Unions*. London: Cassell and Co.

Stace, H.E., Pettitt, C.W. and Waterston, C.D. 1988. *Natural science collections in Scotland*. Edinburgh: National Museums of Scotland.

Stafford, B. 1984. *Voyage into substance; art, science, nature, and the illustrated travel account, 1760–1840*. Cambridge: MIT Press.

Stanley-Price, N. 2024. George Knox, an Irish politician and mineralogist: death on the road to Naples. *Friends of the Non-Catholic Cemetery in Rome Newsletter* 66: 2.

Steenstrup, K.J.V. 1910. Einleitung und biographische Mitteilungen. In Giesecke, C.L. (ed.) *Mineralogisches Reisejournal über Grönland 1806–1813. Meddelelser om Grønland* 35: 1–37.

Stephens, W. and Fitton, W.H. 1812. *Notes on the mineralogy of part of the vicinity of Dublin*. London: William Phillips.

Stewart, A. 1987. *Royal Hibernian academy of arts. Index of exhibitioners 1826–1979. Volume 3, N–Z.* London: Manton Publishing.

Stewart, A. 1995. *Irish art loan exhibitions, 1765–1927. Volume 2, Index of Artists, N–Z.* Jersey: C.I. Appleby.

Stewart, B. and Cutten, M. 1997. *The dictionary of portrait painters in Britain up to 1920*. Woodbridge: Antique Collectors' Club.

Stewart, D. 1799. The report of Donald Stewart, itinerant mineralogist to the Dublin Society. *Transactions of the Dublin Society* 1: 1–142.

Stinner, A. 2002. Calculating the age of the Earth and the Sun. *Physics Education* 37: 296–305.

Stokes, W. 1807. *A catalogue of the minerals in the museum of Trinity College*. Dublin: W. Watson.

Stokes, W. 1818. *A descriptive catalogue of the minerals in the systematic collection of the museum of Trinity College Dublin*. Dublin: University Press.

Strange J. 1775. An account of two Giants Causeways, or groups of prismatic basaltine columns, and other curious volcanic concretions, in the Venetian State in Italy. *Philosophical Transactions of the Royal Society of London* 65: 5–47.

Stronach, G. 1909. Alexander Nimmo. In Lee, S. (ed.) *Dictionary of National Biography*, Volume 14. London: Smith and Elder, 512.

Sweet, J.M. 1967. Robert Jameson's Irish Journal, 1797. *Annals of Science* 23: 97–126.

Taquet, P. 2007. On camelback: René Chudeau (1864–1921), Conrad Kilian (1898–1950), Albert Félix de Lapparent (1905–1975) and Théodore Monod (1902–2000), four French geological travellers cross the Sahara. In Wyse Jackson, P.N. (ed.) *Four centuries of geological travel: the search for knowledge on foot, bicycle, sledge and camel.* Geological Society, London, Special Publications 287: 183–190.

Taylor, K.L. 1969. Nicholas Desmarest and geology in the eighteenth century. In Schneer, C.J. (ed.) *Toward a history of geology.* Cambridge, Massachusetts: MIT Press, 339–356.

Taylor, K.L. 2007. Geological travellers in Auvergne, 1751–1800. In Wyse Jackson, P.N. (ed.) *Four centuries of geological travel: the search for knowledge on foot, bicycle, sledge and camel.* Geological Society, London, Special Publications 287: 73–96.

Taylor, M.A. 1990a. The irrationalization of university geological collections. *Geology Today* 6: 9–10.

Taylor, M.A. 1990b. The irrationalization of university geological collections a postscript (no thanks to the UFC). *Geology Today* 6: 54.

Testa, D. 1793a. *Lettera sui pesci fossili di Monte Bolca*. Milano: Imperial Monistero di S. Ambrogio Maggiore.

Testa, D. 1793b. *Seconda lettera su i pesci fossili del Monte Bolca*. Verona: Eredi Moroni.

Testa, D. 1793c. *Terza lettera sui pesci fossili del Monte Bolca*. Venezia.

Thompson, S.M. 1894. The common mussel (*Mytilus edulis*). *Irish Naturalist* 3: 176.

Thompson, S.M. 1900. The supposed occurrence of White Lias at Macedon Pt. *Irish Naturalist* 9: 154.

Thompson, S.M. 1901a. Report of the Geological Committee [1893–96]. *Proceedings of the Belfast Naturalists' Field Club* 4: 115–125, 229–231, 234–235, 302–310, 386–390.

Thompson, S.M. 1901b. On the supposed occurrence of a patch of white Lias or Rhaetic rock on the shore N. of Macedon Point, Belfast Lough. *Proceedings of the Belfast Naturalists' Field Club* 4: 566–569.

[Thompson] Christen, S.M. 1902. Investigations into the glacial drifts of the north-east of Ireland, conducted by the Naturalists' Field Club. *Irish Naturalist* 11: 275–276.

[Thompson] Christen, S.M. 1906. A summary of the [B.N.F.] Club's recent glacial work. *Irish Naturalist* 15: 80.

[Thompson] Christen, S.M. 1910. *Rodolphe Christen: the story of an artist's life, 1859–1906*. London: Longmans, Green & Co.

Thomson, W. 1854. On the mechanical energies of the Solar System. *Philosophical Magazine*, Series 4, 8: 409–430.

Thomson, W. 1861. On the reduction of observations of underground temperature; with application to Professor Forbes' Edinburgh observations, and the continued Calton Hill series. *Transactions of the Royal Society of Edinburgh* 22: 405–427.

Thomson, W. 1862. On the age of the Sun's heat. *Macmillan's Magazine* 5: 288–293.

Thomson, W. 1864. On the secular cooling of the Earth. *Transactions of the Royal Society of Edinburgh* 23: 157–170.

Thomson, W. 1868. On geological time. *Transactions of the Geological Society of Glasgow* 3: 1–28.

Thomson, W. 1884. On the mechanical energies of the Solar System. *Mathematical and Physical Papers* Volume 2: 1–27. London: Macmillan & Co.

Thomson, W. 1889. On the age of the Sun's heat. *Popular Lectures and Addresses* 1. Cambridge: Cambridge University Press, 349–368.

Thomson, W. 1890. On the secular cooling of the Earth. *Mathematical and Physical Papers* Volume 3: 295–311. London: Macmillan & Co.

Thomson, W. [Lord Kelvin] 1899. The age of the earth as an abode fitted for life. *Science* 9: 665–674; 704–711 [Also published in *Annual Report of the Smithsonian Institution for 1897*, (1898), 337–357; *Journal of the Transactions of the Victoria Institution* 31 (1899): 11–34; and in the *London, Edinburgh, and Dublin Philosophical Magazine and Journal of Science*, Series 5, 57 (1899): 66–90].

Todd, J.H. 1869. *A catalogue of graduates who have proceeded to degrees in the University of Dublin*. Dublin: Hodges, Smith and Foster.

Tomkeieff, S.I. 1983. *Dictionary of petrology*. Chichester and New York: John Wiley and Sons.

Topley, W. 1872. Geology of the straits of dover. *Quarterly Journal of Science* 2: 208–223.

Torrens, H.S. 1985. Early collecting in the field of geology. In Impey, O. and MacGregor, A. (eds.) *The origins of museums*. Oxford: Oxford University Press, 204–213.

Torrens, H.S. 1994. Patronage and problems: banks and the Earth sciences. In Banks, R.E.R., Elliott, B. and Hawkes, J.G. (eds.), *Sir Joseph Banks: a global perspective*. Kew: Royal Botanic Gardens, 49–75.

Torrens, H.S. 1995. Presidential address. Mary Anning (1799–1847) of Lyme; 'the greatest fossilist the world ever knew'. *British Journal for the History of Science* 28: 257–284.

Torrens, H.S. 2001. Timeless order: William Smith (1769–1839) and the search for raw materials. In Lewis, C.L.E. and Knell, S.J. (eds.). *The age of the Earth from 4004 BC to AD2002*. Geological Society, London, Special Publications 190: 61–83.

Torrens, H.S. 2003. An introduction to The Life and Times of William Smith (1769–1839). In Phillips, J. (ed.) *Memoirs of William Smith, LL.D., author of the "Map of the Strata of England and Wales"*. Bath: Bath Royal Literary and Scientific Institution, xi–xxxviii. [New edition of Phillips 1844].

Torrens, H.S. 2004a. The water-related work of William Smith (1769–1839). In Mather, J.D. (ed.) *200 years of British hydrogeology*. Geological Society, London, Special Publications 225: 15–30.

Torrens, H.S. 2004b. The life and work of the geologist S.S. Buckman. *Proceedings of the Cotteswold Naturalists' Field Club* 43: 25–47.

Torrens, H.S. and Taylor, M.A. 1990. Geological collections and museums in Cheltenham 1810–1988: a case history and its lessons. *The Geological Curator* 5: 175–213.

Traxler-Brown, B. 1991. German life and letters as witnessed by the books of an Irish chemist, Dr. Richard Kirwan (1733–1812). In Plassmann, E. Schmitz, W. and Vodosek, P. (eds.) *Buch und Bibliothekswissenschaft im Informationszeilalter.* Munich: G.G. Saur, 297–312.

Tunnicliff, S.P. 1980. *A catalogue of the lower Palaeozoic fossils in the collection of major general J.E. Portlock, R.E., LL.D., F.R.S. &c.* Belfast: Ulster Museum.

Turner, B.S. 1979. *A list of photographs in the R.J. Welch collection in the Ulster Museum, volume 1: topography and history.* Belfast: Ulster Museum.

Vaccari, E. 1993. *Giovanni Arduino (1714–1795). Il contributo di uno scienziato veneto al di-battito settecentesco sulle scienze della Terra.* Firenze: Leo S. Olschki.

Vaccari, E. and Monaghan, N.T. 1993. I minerali di Giovanni Arduino nella collezione geo-mineralogica di Nathanael Gottfried Leske: verifica di un caso di comunicazione scientifica nell Europa del tardo settecento. *Geologica Romana* 29: 547–565.

Vaccari, E. and Wyse Jackson, P.N. 1995. The fossil fishes of Bolca and the travels in Italy of the Irish Cleric George Graydon in 1791. *Museologia Scientifica* 4: 57–81.

Vallance, T.G. 1978. Pioneers and leaders – a record of Australian palaeontology in the nineteenth century. *Alcheringa* 2: 243–250.

Vaughan, W.E. 1980. Richard Griffith and the tenement valuation. In Herries Davies, G.L. and Mollan, R.C. (eds.) *Richard Griffith: 1784–1878.* Dublin: Royal Dublin Society, 103–122.

Villiers-Tuthill, K. 2006. *Alexander Nimmo & the western district.* Clifden: Connemara Girl Publications.

Volta G.S., 1789. *Degl'impietrimenti del territorio Veronese ed in particolare dei pesci fossili del celebre Monte Bolca per servire di continuazione all'argomento delle rivoluzioni terracquee. Lettera [...1 al].* Mantova: Sig. Vincenzo Bozza.

Volta G.S., 1794. *Dei pesci fossili del Veronese. Lettera indirizzata al sig. abate Domenico Testa.* Mantova: G. Braglia.

Walcott, C.D. 1893. Geologic time, as indicated by the sedimentary rocks of North America. *Journal of Geology* 1: 639–676.

Waldie, J. 1859. *A catalogue of pictures, statues, busts, antique columns, bronzes, fragments of antique buildings, tables of Florentine and Roman mosaic, scagliola and inlaid wood; Indian, Neapolitan and other china, with notices of the large collection of books in the various apartments, at Hendersyde Park, to which is added some particulars of the exterior of the house and adjoining buildings, and of the pleasure grounds, gardens, walks, shrubberies, and woods, and of the island of Sharpitlaw, with its woods and walks, and the suspension bridge leading from the walk on the North side of the mill stream to the island.* 2nd edition. Kelso: printed by Robert Stewart for private circulation.

Walker, J. 1793. Memoir upon the marble quarries of Carrara & upon the Ichthyolites of Monte Bolca in the Veronese. (Soc. Prom. Nat. Hist. Manuscript, 7 leaves).

Walley, G.P. (ed.) 1993. *Register of natural science collections in the Midlands of England.* Nottingham: Nottingham City Museums for the West Midlands Area Museum Service and the East Midlands Museums Service.

Waterhouse, G. 1933. Goethe, Giesecke, and Dublin. *Proceedings of the Royal Irish Academy* 41C: 210–218.

Waterhouse, G. 1936. Sir Charles Giesecke's autograph albums. *Proceedings of the Royal Irish Academy* 43C: 291–306.

Watts, W.W. 1898. *Geology for beginners*. London: Macmillan.

Weaver, T. 1819. Memoir of the geological relations of the east of Ireland. *Transactions of the Geological Society of London* 5: 117–304.

Werner, A.G. 1814. *Nomenclature of colours*. Arranged by P. Syme. Edinburgh and London: William Blackwood and T. Cadell [A combination of this edition and the 1821 second edition published 2017. London: Natural History Museum]

Werner, A.G. 1817. *Abraham Gottlob Werners Letztes Mineral System*. Freiberg and Vienna.

White, T. de V. 1955. *The story of the Royal Dublin Society*. Tralee: The Kerryman.

Whitehurst, J. 1786. *An inquiry into the original state and formation of the Earth; deduced from facts and the laws of nature*. 2nd edition. London: printed for W. Bent.

Whittaker, A. 1991. The Magic Flute cast: geological correlations with Mozart. *Terra Nova* 3: 9–16.

Whittaker, A. 2007. The travels and travails of Sir Charles Lewis Giesecke. In Wyse Jackson, P.N. (ed.) *Four centuries of geological travel: the search for knowledge on foot, bicycle, sledge and camel*. Geological Society, London, Special Publications 287: 149–160.

Wigley, P., Darrell, J., Clements, D. and Torrens, H.S. 2019. *William Smith's fossils reunited*. Wellington, Somerset: Halsgrove Publishing.

Wigley, P., Dolan, P., Sharpe, T. and Torrens, H.S. 2007. *Strata Smith: his 200 year legacy, digitally enhanced maps & sections by William Smith, George Bellas Greenough, John Cary & Richard Thomas 1796–1840. CD*. London: Geological Society.

Wilkins, N.P. 2016. *Alexander Nimmo, master engineer, 1783–1932: public works and civil surveys*. Newbridge: Irish Academic Press.

Williams, A. 1973. John Kaye Charlesworth C.B.E., Hon. D.Sc. (Self.), D.Sc. (Leeds), Ph.D (Breslau), M.R.I.A., F.R.G.S., F.G.S. *Royal Society of Edinburgh Year Book 1972–73* (unpaginated).

Williams, R.B. 2017. The palaeontologist William Hellier Baily (1819–1888): new biographical information. *The Geological Curator* 10: 465–467.

Williamson, G.C. 1905. *Bryan's dictionary of painters and engravers, Volume 5, S–Z*. London: George Bell.

Wills, R.G. 1935. The contribution of British medical men to the foundation of geology. *Proceedings of the Liverpool Geological Society* 16: 199–219.

Wilson, H.E. 1972. John Kaye Charlesworth, 1889–1972. *Irish Naturalists' Journal* 17: 209–210.

Wilson, H.E. 1985. *Down to Earth; one hundred and fifty years of the British Geological Survey*. Edinburgh: Scottish Academic Press.

Wilson, J. 1987. Joseph Wright, F.G.S., 1834–1923. *Irish Naturalists' Journal* 22: 169–180.

Wilson, L.G 1988. *Lyell in America. Transatlantic geology, 1841–1853*. Baltimore, MD and London: John Hopkins University Press.

Wilson, W.E. 1994. The history of mineral collecting. *Mineralogical Record* 25: 1–243.

Woodward, H.B. 1907. *The history of the Geological Society of London*. London: Longman, Green and Co.

Worth, R.H. 1908. The dredging of the Marine Biological Association (1895–1906), as a contribution to the knowledge of the geology of the English Channel. *Journal of the Marine Biological Association of the United Kingdom* 8: 118–188.

Wright, E.P. and Huxley, T.H. 1866. On a collection of fossils from the Jarrow Colliery, Kilkenny, Ireland. *Geological Magazine* 3: 165–171.

Wright, G.N. 1821. *An historical guide to ancient and modern Dublin*. London: Baldwin, Cradock and Joy.

Wright, J. in Greenly, E. 1900. Report on the drift at Moel Tryfaen. *Geological Magazine* 7: 115–123.

Wright, T. 1855. On fossil Echinoderms from the island of Malta, with notes on the stratigraphic distribution of the fossil organisms in the Maltese Beds. *Annals and Magazine of Natural History*, Series 2, 15: 101–127; 175–196; 262–276.

Wyllie, P.J. 1998. Hutton and Hall on theory and experiments: the view after 2 centuries. *Episodes* 21: 3–10.

Wyse Jackson, P.N. 1987. Statistical survey of geological and geomorphological excursions undertaken by the Dublin Naturalists' Field Club 1886–1986. *Dublin Naturalists' Field Club Occasional Publication* 4: 1–14.

Wyse Jackson, P.N. 1989. On rocks and bicycles: a biobibliography of Grenville Arthur James Cole (1859–1924), fifth Director of the Geological Survey of Ireland. *Geological Survey of Ireland Bulletin* 4: 151–163.

Wyse Jackson, P.N. 1991. The cycling geologist. *Cycle Touring and Campaigning* 4 (June–July): 26–27.

Wyse Jackson, P.N. 1992a. A man of invention: John Joly (1857–1933), engineer, physicist, and geologist. In Scott, D.S. (ed.) *Treasures of the mind: a Trinity College Dublin quatercentenary exhibition*. London: Sothebys, 89–96, 158–160.

Wyse Jackson, P.N. 1992b. The geological collections of Trinity College, Dublin. *The Geological Curator* 5: 263–74.

Wyse Jackson, P.N. 1994a. Fish and other fossils from the Eocene of Bolca, Italy. *The Geological Curator* 6: 103–105.

Wyse Jackson, P.N. (ed.) 1994b. *In marble halls: Geology in Trinity College, Dublin*. Dublin: Department of Geology, Trinity College.

Wyse Jackson, P.N. 1994c. A history of geology in Trinity College, Dublin. In Wyse Jackson, P.N. (ed.) *In marble halls: Geology in Trinity College, Dublin*. Dublin: Department of Geology, Trinity College, 13–78.

Wyse Jackson, P.N. 1995. Patrick Ganly (1809–1899) and the discovery of evidence of way-up in rocks on the Dingle Peninsula. *The Kerry Magazine* 7: 8–9.

Wyse Jackson, P.N. 1996a. Alexander Nimmo's *On the Application of the science of Geology to the purposes of Practical Navigation* (1825): the first investigation of marine geology and its bearing on the geology of offshore south west British Isles. *Earth Sciences History* 15: 167–171.

Wyse Jackson, P.N. 1996b. Sir Charles Lewis Giesecke (1761–1833) and Greenland: a recently discovered mineral collection in Trinity College, Dublin. *Irish Journal of Earth Sciences* 15: 161–168.

Wyse Jackson, P.N. 1997a. Fluctuations in fortunes: geology in Ireland 1694–1994. In Wilson Foster, J. and Chesney, H.G.C. (eds.) *Nature in Ireland*. Dublin: Lilliput Press.

Wyse Jackson, P.N. 1997b. John W. Pringle (*c*. 1790–1861) and ordnance survey geological mapping in Ireland. *Proceedings of the Geologists' Association* 108, 153–156.

Wyse Jackson, P.N. 1998a. William Henry Fitton (1780–1861) and the Wollaston Medal of 1852. *Geoscientist* 8(10): 9.

Wyse Jackson, P.N. 1998b. Hibernian research on Italian volcanic rocks: the Mallet and Guiscardi/Haughton collections in Trinity College, Dublin, Ireland. In Morello, N. (ed.)

Volcanoes and History. Proceedings of the 20th INHIGEO symposium, Naples-Eolie-Catania, 19–25 September 1995. Genoa: Brigati, 675–685.

Wyse Jackson, P.N. 1998c. Richard Kirwan: chemist and geologist. *Irish Chemical News* 12: 36–39.

Wyse Jackson, P.N. 1998d. Veronica Burns 1914–1998, Curator Geological Museum, Trinity College, Dublin 1964–1980. *The Geological Curator* 6: 339–340.

Wyse Jackson, P.N. 1998e. Veronica Conroy Burns (1914–1998). *Irish Naturalists' Journal* 26: 1–4.

Wyse Jackson, P.N. 2000. Tumultuous times: geology in Ireland and the debate on the nature of basalt and other rocks of north-east Ireland between 1740 and 1816. In Wyse Jackson, P.N. (ed.) *Science and engineering in Ireland in 1798: a time of revolution*. Dublin: Royal Irish Academy, 35–49.

Wyse Jackson, P.N. 2001a. John Joly (1857–1933) and his determinations of the age of the Earth. In Lewis, C.L.E. and Knell, S.J. (eds.) *The age of the Earth: from 4004 BC to AD 2002*. Geological Society, London, Special Publications 190: 107–119.

Wyse Jackson, P.N. 2001b. Chronological listing of books published on the geology of Ireland from 1699 to 1999. *Irish Naturalists' Journal* 26: 293–313.

Wyse Jackson, P.N. 2002a. Classic paper in the history of geology: John Joly's paper: "Uranium and Geology" (1908). *Episodes* 25: 258–263.

Wyse Jackson, P.N. 2002b. Henry Benedict Medlicott, geologist 1829–1905. In Mollan, R.C., Davis, W.J. and Finucane, B. (eds.), *Irish innovators in science and technology*. Dublin: Royal Irish Academy, 144.

Wyse Jackson, P.N. 2002c. Richard Dixon Oldham, geologist and seismologist, 1858–1936. In Mollan, R.C., Davis, W.J. and Finucane, B. (eds.), *Irish innovators in science and technology*. Dublin: Royal Irish Academy, 187.

Wyse Jackson, P.N. 2004. Thomas Hawkins, Lord Cole, William Sollas and all: casts of Lower Jurassic marine reptiles in the Geological Museum, Trinity College, Dublin, Ireland. *The Geological Curator* 8: 11–18.

Wyse Jackson, P.N. 2006a. *The Chronologers' Quest: episodes in the search for the age of the Earth*. Cambridge: Cambridge University Press.

Wyse Jackson, P.N. 2006b. The Honorable George Knox (1765–1827), parliamentarian and mineral collector: his collection in Trinity College, Dublin, Ireland. *The Mineralogical Record* 37: 543–551.

Wyse Jackson, P.N. (ed.) 2007a. *Four centuries of geological travel: the search for knowledge on foot, bicycle, sledge and camel*. Geological Society, London, Special Publications 287.

Wyse Jackson, P.N. 2007b. Grenville Arthur James Cole (1859–1924) the cycling geologist. In Wyse Jackson, P.N. (ed.) *Four centuries of geological travel: the search for knowledge on foot, bicycle, sledge and camel*. Geological Society, London, Special Publications 287: 135–148.

Wyse Jackson, P.N. 2007c. Irish "rock stars": John Joly (1857–1933). *Earth Science Ireland* 2: 11.

Wyse Jackson, P.N. 2007d. Catalogue of lavas di Robert Mallet. In Ferrari, G. (ed.) *Viaggio nelle aree del terremoto del 16 dicembre 1857, Volume 2, Part 4. Laboratorio Mallet. Fonti e studi su Robert Mallet e sui caratteri geodinamici dell'area lucana*. Bologna: SGA Storia Geofisica Ambiente, 95–102.

Wyse Jackson, P.N. 2008a. Sir Richard John Griffith (1784–1878): a portrait in oils by Stephen Catterson Smith (1806–1872). *Irish Journal of Earth Sciences* 26: 45–51.

Wyse Jackson, P.N. 2008b. Servants of the Raj at play: an example of the social interactions between a geologist and an engineer in nineteenth century India. *The Compass* 81: 96–103

Wyse Jackson, P.N. 2009a. Erratics, intrusions and graptolites. In Mulvihill, M. (ed.) *Lab Coats and Lace: the lives and legacies of inspiring Irish women scientists and pioneers.* Dublin: WITS, 48–59.

Wyse Jackson, P.N. 2009b. Irish 'Rock Stars': Charles Irving Gardiner (1868–1940) & Sydney Hugh Reynolds (1867–1949). *Earth Science Ireland* 6 (2009), 12–13.

Wyse Jackson, P.N. 2011. History of ichnology: John Joly (1857–1933) on *Oldhamia*: poetic and scientific observations. *Ichnos* 18: 209–211.

Wyse Jackson, P.N. 2016. William Smith and Ireland: sources of Irish geological information on his geological map of 1820. *Earth Sciences History* 35, 115–123.

Wyse Jackson, P.N. 2019a. Irish 'Rock Stars': John Hart (1797–1872). *Earth Sciences Ireland* 24: 8–9.

Wyse Jackson, P.N. 2019b. Mary Katherine Andrews (1851–1914). In Dalton, C. and McGlynn, G. (eds.) *Giants of the Irish Quaternary.* Dublin: Irish Quaternary Association, 37–43.

Wyse Jackson, P.N. 2019c. Vertical and horizontal: changes in the design, settings and use of the Museum Building 1853–2018. In Casey, C. and Wyse Jackson, P.N. (eds.) *The Museum Building of Trinity College Dublin: setting the standard in design, materials and craftsmanship.* Dublin: Four Courts Press, 304–341.

Wyse Jackson, P.N. and Caulfield, L.M. 2023. The rough and the smooth: stone use in Dublin 1720–1760. In Casey, C. and Hayes, M. (eds.). *Enriching architecture: craft and its conservation in Anglo-Irish building production, 1660–1760.* London: UCL Press, 234–261.

Wyse Jackson, P.N. and Hawley, D. 2023. Richard Griffith's small scale geological map of Ireland (1853). *INHIGEO Anniversaries, IUGS E-Bulletin* 199: 1–5. (https://www.inhigeo.com/anniversaries/Wyse_Jackson,_P._N.,_Hawley,_D._Griffith's_Map_of_Ireland_170_years_ago.pdf)

Wyse Jackson, P.N. and Monaghan, N.T. 1994. Frederick M'Coy (*c.* 1823–1899): an eminent Victorian palaeontologist and his synopses of Irish palaeontology of 1844 and 1846. *Geology Today* 10: 231–234.

Wyse Jackson, P.N. and Monaghan, N.T. 1995. Transfer of the Huxley and Wright (1867) Carboniferous amphibian and fish material to Trinity College Dublin from the National Museum of Ireland. *Journal of Paleontology* 69: 602–603.

Wyse Jackson, P.N. and Parkes, M.A. 2009. William Hellier Baily (1819–1888): forever an Acting Palaeontologist with the Geological Survey of Ireland. *The Geological Curator* 9: 57–84.

Wyse Jackson, P.N. and Parkes, M.A. 2017. Burial place and headstone of William Hellier Baily (1819–1888). *The Geological Curator* 10: 375.

Wyse Jackson, P.N. and Spencer Jones, M.E. 2007. The quiet workforce: the various roles of women in geological and natural history museums during the early to mid-1900s. In Burek, C.V. and Higgs, B. (eds.) *The role of women in the history of geology.* Geological Society, London, Special Publications 281: 97–113.

Wyse Jackson, P.N. and Vaccari, E. 1993. Volcanoes and Straw Bonnets: The Graydons of Burrishoole. *Cathair na Mart: Journal of the Westport Historical Society* 13: 90–101.

Wyse Jackson, P.N. and Vaccari, E. 1997. *The Reverend George Graydon (c. 1753–1803): cleric and geological traveller.* Dublin: Royal Irish Academy.

Wyse Jackson, P.N. and Wyse Jackson, P.S. 1992. *The Irish Naturalist*: 33 years of natural history in Ireland 1892–1924. *Irish Naturalists' Journal* 24: 95–101.

Wyse Jackson, P.N. and Wyse Jackson, P.S. 2019. A stone menagerie and botanical conservatory: the naturalistic carvings of the Museum Building. In Casey, C. and WyseJackson, P.N. (eds.) *The Museum Building of Trinity College Dublin: setting the standard in design, materials and craftsmanship.* Dublin: Four Courts Press, 216–251.

Zammit Maempel, G. 2007. *Illustrators and their illustrations of Maltese fossils and Geology: a historical and biographical account.* San Gwann, Malta: Publishers Enterprises Group (PEG) Ltd.

Zwaan, P.C., Petersen, O.V. and Hooker, M. 1974; revised 1977. *World directory of mineral collections.* Leiden: International Mineralogical Association.

INDEX

Note: **Bold** page numbers refer to tables and *italic* page numbers refer to figures.

FSC
www.fsc.org
MIX
Papier | Fördert
gute Waldnutzung
FSC® C083411

Kolibri 360 GmbH
Ferdinand-Jühlke-Straße 7
99095 Erfurt, Deutschland
produktsicherheit@kolibri360.de